EMBEDDED SYSTEM
APPLICATIONS

T0205340

EMBEDDED SYSTEM APPLICATIONS

EDITED BY

Claude BARON
Jean-Claude GEFFROY
Gilles MOTET
Institut National des Sciences Appliquées

Toulouse, France

KLUWER ACADEMIC PUBLISHERS

BOSTON / DORDRECHT / LONDON

A C.I.P. Catalogue record for this book is available from the Library of Congress

ISBN 978-1-4419-5179-3

Published by Kluwer Academic Publishers,
P.O. Box 17, 3300 AA Dordrecht, The Netherlands.

Sold and distributed in the U.S.A. and Canada
by Kluwer Academic Publishers,
101 Philip Drive, Norwell, MA 02061, U.S.A.

In all other countries, sold and distributed
by Kluwer Academic Publishers Group,
P.O. Box 322, 3300 AH Dordrecht, The Netherlands.

Printed on acid-free paper

FOREWORD

ATW'96 EUROPE Organization Board.

Since 1992, the Advanced Technology Workshop (ATW) provides a forum for Engineers, Researchers, and Managers in the US and Europe to exchange concepts and methodologies for the application of advanced technologies to real world applications. This workshop has two parts: the EUROPEAN part and the US part. In 1996, ATW'96 EUROPE (July, 8-10) has been hosted by the Laboratoire d'Etude des Systèmes Informatiques et Automatiques of Institut National des Sciences Appliquées, in Toulouse, France, and ATW'96 US (August, 7-9) has been hosted by the US Air Force, Hanscom Air Force Base, in Bedford, Massachusetts, USA.

ATW'96 EUROPE emphasized Collaborative Engineering applied to Hardware/Software Embedded System Applications. It offered participants coming from the University and Industry several Technical and Panel sessions (Design & Test, Dependability, Spatial Applications, Collaborative Engineering, Education), Invited Speeches and Software Tool Demonstrations. This book collects papers presented in the technical sessions, a synthesis of the Education Panel, and two presentations of software tools. We are particularly grateful to Bernard Salvagnac and Jean-Marie Kubek who had the hard job to convert and integrate the files from various formats in order to produce this text.

The editors

Claude Baron,
Jean-Claude Geffroy,
Gilles Motet.
Laboratoire d'Etude des Systèmes Informatiques et Automatiques, INSAT-DGEI,
Complexe Scientifique de Rangueil, 31077 TOULOUSE Cedex, France.
phone: +33 5 61 55 98 13, fax: +33 5 61 55 98 00, e-mail: lesia@dge.insa-tlse.fr

We wish to thank for their contribution to the success of ATW'96 EUROPE:

- *the United States Air Force, European Office of Aerospace Research and Development,*
- *the région 'Midi Pyrénées',*
- *the INSAT Institute.*

ATW'96 EUROPE Program

Technical Sessions:

Design & Test I	*chair*: A. Rucinski, University of New Hampshire, USA
Design & Test II	*chair*: P. Sanchez, University of Cantabria, Spain
Dependability	*chair*: R. Straitt, US Air Force, USA
Spatial Applications	*chair*: P. David, Matra Marconi Space, France
Collaborative Engineering	*chair*: J-F. Santucci, University of Corte, France

Panel Sessions:

Dependability and Design Methods,
    moderator: D. Boutin, Lockheed Martin Aeronautical Systems, USA
Education "Use of MultiMedia and Decentralized Means in Education",
    moderator: J-P. Soula, CERDIC-INSAT, France
Collaborative Engineering "Industry & Research Cooperation",
    moderators: J-F. Santucci & M. Filippi, University of Corte, France

Invited Speeches:

. *The Challenge of Future Decision Support,*
Col. R. Adams, ESC System Integration Office, Hanscom AFB, USA
. *USAF International Cooperative Projects,*
B.J. Durante, US Air Force, Pentagon, USA
. *Developments of Space Qualified Digital Correlators:*
Spectrometers for the European Sub-millimiter Space Telescope (FIRST Project),
M. Giard, CESR-CNRS, France.

ATW'96 EUROPE Steering Committee:

C. Baron, LESIA INSA, France,
D. Boutin, Lockheed Martin, USA,
J-C. Geffroy, LESIA INSAT, France,
G. Motet, LESIA INSAT, France,
A. Rucinski, University of New Hampshire, USA,
J.-F. Santucci, University of Corte, France,
R. Straitt, US Air Force, USA

ATW'96 US - EUROPE Program Committee:

General chair:
J.-F. Santucci, University of Corte, France,
Co-chairs:
C. Baron, INSA Toulouse, France (chair of the European part),
R. Straitt, US Air Force, USA (chair of the US part)
Program chairs:
B. Dziurla-Rucinska, University of New Hampshire, USA,
P. Sanchez, University of Cantabria, Spain,
Other Members:
P. Arato, Technical University of Budapest, Hungary,
P. Bisgambiglia, University of Corte, France,
W. Ellis, Software Process and Metrics, USA,
D. Erbschloe, US Air Force, USA,
M. Filippi, University of Corte, France,
D. Forest, University of New Hampshire, EOS, USA,
J.C. Geffroy, LESIA, INSA Toulouse, France,
P. Girard, University of Montpellier, France,
J. Gueller, Texas Instrument, USA,
R. Hermida, University of Complutense, Spain,
D. Hummer, University of Duisbourg, Germany,
W. Jaworski, Concordia University, Canada,
H. Joiner, SRC, USA,
G. Motet, LESIA, INSA Toulouse, France,
A. Rucinski, University of New Hampshire, USA,
B. Straube, FhGIIS, Germany,
T. Tewksbury, University of West Virginia, USA.

CONTENTS

7 EVALUATION OF AN INTEGRATED HIGH-LEVEL SYNTHESIS METHOD

8 COMBINATORIAL CRITERIA OVER GRAPHS OF SPECIFICATION TO DECIDE SYNTHESIS BY SEQUENTIAL CIRCUITS

9 AUTOMATIC GENERATION AND OPTIMISATION OF MARKOV MATRICES

10 FAULT MODELING IN SPACE-BORNE RECONFIGURABLE MICROELECTRONIC SYSTEMS

THE JOINT SYSTEMS/SOFTWARE ENGINEERING ENVIRONMENT (JOSEE) CONCEPT AT LOCKHEED MARTIN AERONAUTICAL SYSTEMS

D. R. Boutin

Software Engineering Process Department
Lockheed Martin Aeronautical Systems, Dept. 73-F9/Zone 0685
86 South Cobb Drive, Marietta, GA 30063-0685, USA
Phone: (404) 494-9634, Fax (404) 494-1661,
e-mail: dboutin@mantis.mar.lmco.com

ABSTRACT

We can define Systems/Software Engineering Environments (S/SEEs) as being "A set of software-based tools that aid in providing total integrated lifecycle support for systems and software development, augment all test activities and aid in the management of projects and programs." In other words, an integrated series of software programs that provide partial or total automation of the activities within system and software lifecycles.

Ideally, a S/SEE would be convenient to use, support customization of data and integrated toolsets, have an open architecture, support the selected software development process methodologies, encompass the entire database while providing tool interfacing and evolution and support standards which enable portability and interoperability.

S/SEEs, though a "hot topic", are a fairly new idea in software development. Until recently, when computer-aided software development was thought of, one would think of individual Computer Aided Software Engineering (CASE) development tools. One tool, one lifecycle phase. Only since the late-1980s has the idea of a fully integrated S/SEE been discussed or proposed-by vendors and users alike.

1 LOCKHEED MARTIN AERONAUTICAL SYSTEMS (LMAS) S/SEE: JOSEE

At Lockheed Martin Aeronautical Systems (LMAS), we have evolved our software development process from the primitive end of the software development lifecycle-no tool automation, to the more sophisticated individual CASE tools, both upper-CASE and lower-CASE, and onward to partially integrated S/SEEs.

LMAS has developed a concept, the Joint S/SEE, or JOSEE, that is a process-centered automated workflow environment which combines the framework technologies and components from several CASE tool and S/SEE technology vendors. The "J" in the acronym "JOSEE" stands for both the "joint" cooperation and integration of S/SEE vendors and CASE tool vendors into a common technology based framework, and for the "joint" cooperation between several different Lockheed Martin companies which support the concept of a common, open, integrated, heterogeneous S/SEE.

The following paragraphs discuss the requirements, vision, capabilities, management and process-methods-tools integration of the JOSEE.

The JOSEE vision includes providing an affordable, consistent, real-time view of the software development process. This ensures that the software manager and project manager perform pro-active rather than re-active software management. The process instantiation and sophisticated metrics provisions and integrated software management capabilities make the JOSEE solution attractive to managers and developers alike at LMAS.

The JOSEE goal is to reduce the cost of software by providing automated means to perform previously manual activities, and to provide management with a sophisticated method of tracking software project status.

2 THE JOSEE COMMON FRAMEWORK

The JOSEE is based upon open industry standards and technologies. The JOSEE framework consists of the Digital Equipment Corporation COHESION product front-end and desktop. This S/SEE framework product supports the Common Object Request Broker Architecture (CORBA) framework technology and also provides support for the Open Software Foundation (OSF) Distributed

Computing Environment (DCE) middleware. These technologies provide a means for support of distributed tooling, licensing and databases and support for client/server technologies providing access to over 21 platforms.

JOSEE: **Joint LMAS/LTAS S/SEE**

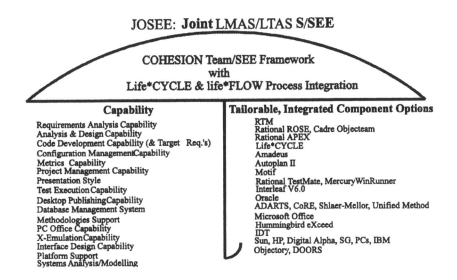

COHESION Team/SEE Framework
with
Life*CYCLE & life*FLOW Process Integration

Capability	Tailorable, Integrated Component Options
Requirements Analysis Capability	RTM
Analysis & Design Capability	Rational ROSE, Cadre Objecteam
Code Development Capability (& Target Req.'s)	Rational APEX
Configuration ManagementCapability	Life*CYCLE
Metrics Capability	Amadeus
Project Management Capability	Autoplan II
Presentation Style	Motif
Test Execution Capability	Rational TestMate, MercuryWinRunner
Desktop PublishingCapability	Interleaf V6.0
Database Management System	Oracle
Methodologies Support	ADARTS, CoRE, Shlaer-Mellor, Unified Method
PC Office Capability	Microsoft Office
X-EmulationCapability	Hummingbird eXceed
Interface Design Capability	IDT
Platform Support	Sun, HP, Digital Alpha, SG, PCs, IBM
Systems Analysis/Modelling	Objectory, DOORS

Figure 1 The JOSEE Solution

The COHESION desktop is a Motif-compliant desktop which provides a common "look and feel" for the desktop across all platforms for which it can be displayed. Figure 1 shows the capabilities and their associated solution sets for the JOSEE.

3   LMAS JOSEE: AUTOMATED PROCESS SUPPORT

The JOSEE utilizes Computer Resources International's (CRI) Life*FLOW workflow automation tool to instantiate the LMAS Standard Software Process (SSP) in an automated fashion which is invisible to the user/developer. The LMAS SSP is based upon ISO 12207 and incorporates attributes from ISO

9000-3, DO/178-B and the Software Engineering Institute's (SEI) Capability
Maturity Model (CMM). The Life*FLOW instantiation provides integrated
tooling automation for all areas of the software development effort. Examples
include automated peer review notification and automated metrics collection.

The JOSEE SSP is modeled using the IDEF0 notation and tool suite. The
COHESION desktop allows for browsing of this model using the Netscape Navigator Version 2.0. The IDEF0 models are linked to the actual Life*FLOW instantiated process using the Netscape Navigator and JAVA technologies. From
within COHESION a user may browse the SSP on the JOSEE home page and
link up with more sophisticated on-line help and process assets. As an example, a user may wish to view the requirements management process in IDEF0,
then drill down to the Life*FLOW instantiation inside the JOSEE workflow
automation tool, view some data using the Requirements Traceability Matrix
(RTM) tool, and finally "hot link" to the LMAS SEPD home page which contains the on-line policies/procedures and "how-to" guidebooks for requirements
management. In summary, with a few mouse clicks, a user can get on-line tools
help, process help and LMAS policies and procedures and view assets without
ever having to leave his/her personal computer or workstation.

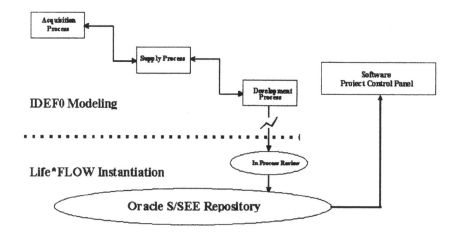

Figure 2   The Process Modeling/Instantiation Relationship

Figure 2 shows the relationship between the JOSEE process modeling capabilities and the process instantiation inside the JOSEE and its subsequent display on the Software Project Control Panel (SPCP), which is described in the following paragraphs.

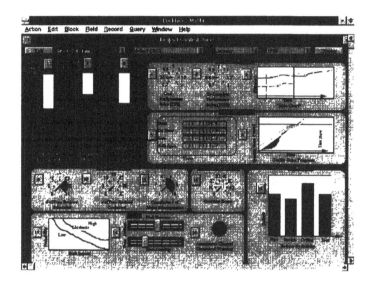

Figure 3   The Software Project Control Panel

4   LMAS JOSEE: INTEGRATED SOFTWARE MANAGEMENT

The LMAS JOSEE includes an integrated software management system designed around the Software Project Control Panel (SPCP), as shown in Figure 3. The SPCP provides a graphical user interface (GUI) view into the current state of the software development project. The view of the SPCP provided to the user is linked to the Life*FLOW defined user role, such as developer, manager, quality assurance (QA) or vice president of engineering. The SPCP is flexible enough to be customized by each user, or according to a project defined ruleset. The SPCP GUI provides a palette of gauges which are "point and click" accessible for each user to customize his/her own SPCP for their desired personal view. Each gauge represents a process or product metric inside an LMAS developed metrics repository. This metrics repository provides an open application programming interface (API) to enable other third party CASE tools to insert data for display onto an appropriate gauge on the SPCP. As an example, the Requirements Traceability Matrix (RTM) Tool from GEC-Marconi can provide requirements database information the LMAS metrics repository via a SQL-Net interface between the RTM Oracle database and the LMAS Oracle metrics repository.

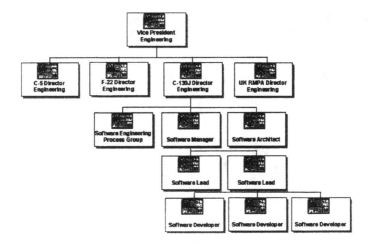

Figure 4

5 LMAS JOSEE: AN INTEGRATED PROGRAM SUPPORT ENVIRONMENT

The LMAS JOSEE provides an integrated approach to a S/SEE. It utilizes the "best of the industry" for the four major components in the S/SEE: (1) Desktop Presentation Integration, (2) Process Integration, (3) Database Integration, and (4) Messaging Integration. The desktop and messaging integration are provided by the Digital Equipment Corporation COHESION product. The process integration and the database integration are provided by the CRI Life*FLOW and Life*CYCLE tools. LMAS has provided its own custom capabilities to tie these two product suites together into a process-centered, automation-based management tool by including the metrics Oracle based metrics repository which provides the capability to link the databases of the various tools for metrics presentation on the SPCP.

6 CONCLUSION

The JOSEE solution provides a "state of the art" implementation of "state of the practice" tooling and S/SEE framework technologies. Figures 5 and 6 show the evolution of the S/SEE industry and the future of the JOSEE concept and vision. The ultimate goal remains creating an integrated development environment for hardware and software technologies, such as VHDL and Ada.

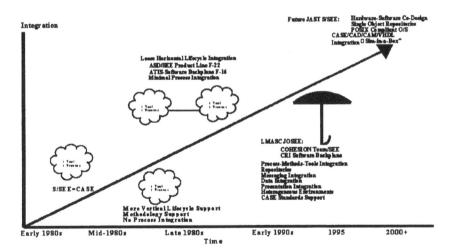

Figure 5 The Evolution of LMAS Environnement

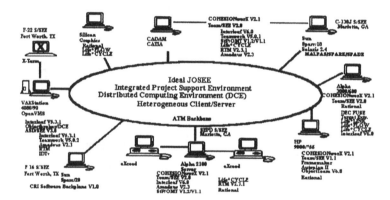

Figure 6

2

CURRENT MODELING IN VITAL

J.L. Barreda, P. Sánchez

Microelectronics Group. TEISA. ETSIIyT
University of Cantabria. Santander. Spain
Ph: 34-42-201548 Fax: 34-42-201873
e-mail: sanchez@teisa.unican.es

ABSTRACT

The recently approved VITAL [7] standard will permit many of the drawbacks presented by the use of VHDL at gate level to be overcome. It does not, however, address one of the basic problems at gate level: current modeling.

The main purpose of the present work is to propose a current modeling for VHDL gate-level descriptions which are VITAL compliant. This technique will be applied to different areas, such as low power design, BIST scheduling and fault simulation, for current fault modeling and for power estimation and average/peak current determination with a maximum variation of 10% with respect to the data obtained by SPICE LEVEL 3[1]. Logically, the new types, signals and subprograms used in current modeling do not verify the modeling rules of the recently approved VITAL standard, constituting a proposal for a possible extension in the future.

The order of contents of this paper will be as follows. In the next section the concepts necessary for transitory current modeling will be introduced. Then, an example of the application of this technique will be presented. The final section will present the conclusions of the article.

1 CURRENT MODELING

With the aim of developing current models for VITAL cells some non-VITAL standard types and subprograms have been introduced.

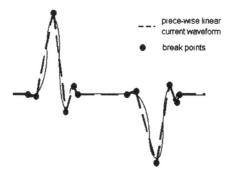

Figure 1   Current waveform

In order to be able to define a current modeling in VITAL, it is necessary to model the current flowing from the circuit to the sensor or power supply in a continuous form, considering all behavior to be transitory. This is not possible using an event-driven simulator. For this reason, the current waveform has been modeled by means of a piece-wise linear current waveform (figure 1), in a similar way to [1][2][3]. This enables results to be obtained which are in disagreement with, at most, 10% of those obtained with SPICE [1]. The basic idea is that the current waveforms for any change in the inputs of the VITAL cell are modeled by the foundry as a set of time-current pairs (figure 2).

Comments of figure 2:

$$\text{Input change} \implies \text{current waveform} = \{(0, i\_0), (t\_1, i\_1), (t\_2, i\_2), ..., (t\_6, i\_6)\}$$
$$i = \text{current}$$
$$t = \text{Time}$$

The VHDL simulator must be able to add up all of these piece-wise linear current waveforms to obtain the total current waveform. The new approach introduced in this paper is that the VHDL simulator does not work with current values, but rather with the points at which the slope of the waveform varies (break points). In this way, the total waveform can be calculated with great speed and accuracy (Figure 3).

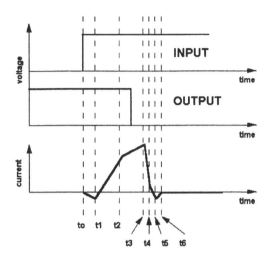

Figure 2

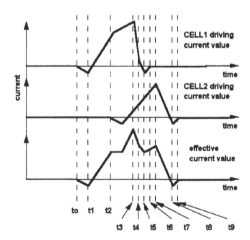

Figure 3

In our approach, the list of pairs presented previously is transformed into a set of linear equations as follows:

Input change $\Longrightarrow$ Current = $\begin{array}{ll} P_1 * T + C_1 & t_0 < T < t_1 \\ P_2 * T + C_2 & t_1 < T < t_2 \\ \qquad ... \\ P_N * T + C_N & t_{n-1} < T < t_n \end{array}$

Where:

T = Continuous Time
$P_j = (i_{j-1} - i_j) / (t_{j-1} - t_j)$
$C_j = i_{j-1} - (P_j * t_{j-1})$

The current is modeled in VHDL by means of resolved signals of record type (Dcurrent) in which one field models the parameter P (slope) and the other the parameter C (initial point) of the linear equation.

```
type current is record
    P : REAL;
    C : REAL;
end record;
type CurrentArray is array( NATURAL range <>) of current;
function DynamicPowerSource( A : CurrentArray ) return current;
subtype Dcurrent is DynamicPowerSource current;
```

Thus, the calculation of the total current is defined as the sum of the currents of each cell. These are modeled by means of equations of the following equation type, in which (P, C) vary as discrete events.

$$Cell_i\text{\_current} = P_i * \text{Time} + C_i$$

Hence, the total current is calculated by means of the following expression :

$$Total\text{\_current} = \Sigma\ Cell_i\text{\_current} = (\Sigma\ P_i\ ) * \text{Time} + (\Sigma\ C_i)$$

This equation is implemented in the resolution function of the Dcurrent type. Thus, this will be the type of signal used to model the power source.

```
function DynamicPowerSource( A : CurrentArray ) return current is
variable result : current := (0.0 , 0.0);
    begin
        IF( A'LENGTH=1) THEN return( A(A'LOW)); ELSE
            for i in A'RANGE loop
                result.P := result.P + A(i).P;
                result.C := result.C + A(i).C;
            end loop;
        END IF;
return result;
end;
```

The advantage of this type is that it allows the current waveform to be modeled with relative accuracy. From the point of view of VITAL, a new port (of the Dcurrent type) is introduced in the cells which models the dynamic current consumption of the cell. Also, a generic port is added, enabling the time-P-C sets to be specified by means of the VitalCurrentTableType type. Every row of this table is associated to the transaction modeled in the same row of a VitalStateTableType constant. These tables are processed inside the cell by means of a VitalDynamicCurrent subprogram. The types necessary for the simulation and the VitalDynamicCurrent subprogram are shown below:

```
type VitalCurrent is record
        Point : current;
        T : TIME;
    end record;
type VitalCurrentTableType is array(NATURAL range <>,NATURAL range <>) of VitalCurrent;
procedure VitalDynamicCurrent ( constant CurrentTable : in VitalCurrentTableType;
        constant StateTable : in VitalStateTableType;
        signal DataIn : in std_logic_vector;
        signal Result : out current ) is
    constant InputSize : INTEGER:= DataIn'LENGTH;
    variable DataInAlias : std_logic_vector(0 to InputSize -1);
```

```
variable StateTableAlias : VitalStateTableType(0 TO (StateTable'LENGTH(1)-1),
            0 TO (StateTable'LENGTH(2)-1)) := StateTable;
variable CurrentTableAlias: VitalCurrentTableType(0 TO (CurrentTable'LENGTH(1)-1),
            0 TO (CurrentTable'LENGTH(2)-1)) := CurrentTable;
type EventCurrentTableType is array (Natural range <>) of VitalCurrent;
variable EventCurrentTable : EventCurrentTableType( 0 TO (StateTable'LENGTH(1)-1));
variable maxEventIndex : Natural := 0;
variable nextEventIndex : Natural := 0;
variable Index : INTEGER;
variable currentTime : TIME;
variable nextTime : TIME := TIME'HIGH;
variable nextEvent: TIME;
variable PrevTime : TIME;
variable temp : current;
variable PrevDataIn : Std_logic_vector(0 to DataIn'LENGTH-1):= (others=>'X');
begin
infinite: LOOP -- Infinite loop
if(maxEventIndex /= nextEventIndex) then
            wait on Datain for nextEvent;
        else
            wait on Datain;
end if;
DataInAlias:= To_X01(DataIn);
currentTime := now;
if(DataInAlias/=PrevDatain) then
        if(NextEvent /= TIME'HIGH and nextEventIndex < maxEventIndex) then
            if(nextTime > currentTime) then
                NextEvent := nextTime - currentTime;
else                assert false
                report "Time computation error in VitalDynamicCurrent";
            end if;
        end if;
assert( StateTable'LENGTH(1)=CurrentTable'LENGTH(1))
        report "Incorrect current table";
col_loop: FOR i IN StateTableAlias'RANGE(1) LOOP
- Check each input element of the entry
row_loop  FOR j IN 0 TO InputSize LOOP
        IF (j = inputSize) THEN - This entry matches
            nextEventIndex:=0;
            maxEventIndex:=0;
            nextTime:=currentTime;
            PrevTime:= 0 ns;
```

```
current_loop: for k in 0 to (CurrentTable'LENGTH(2)-1) LOOP
    EventCurrentTable(maxEventIndex).point.A :=
        CurrentTableAlias(i,k).point.A;
    EventCurrentTable(maxEventIndex).point.B :=
        CurrentTableAlias(i,k).point.B -
    (CurrentTableAlias(i,k).point.A
    *(TimeToReal(currentTime)+ TimeToReal(PrevTime)));
    EventCurrentTable(maxEventIndex).T := CurrentTable(i,k).T;
    maxEventIndex := maxEventIndex +1;
    if(CurrentTable(i,k).T = TIME'HIGH) then
        exit current_loop;
    else
        PrevTime:= CurrentTableAlias(i,k).T + PrevTime;
    end if;
end loop current_loop;
exit col_loop;
    end if;
exit row_loop when not
        StateTableMatch(PrevDataIn(j), DataInAlias(j), StateTableAlias(i,j));
end LOOP row_loop;
end LOOP col_loop;
PrevDataIn := DataInAlias;
end if;
if(nextTime=currentTime and maxEventIndex > nextEventIndex) then - update events
    temp.A := EventCurrentTable(nextEventIndex).point.A;
    temp.B := EventCurrentTable(nextEventIndex).point.B;
    if(EventCurrentTable(nextEventIndex).T = TIME'HIGH) then
        nextTime := TIME'HIGH;
        nextEvent:= TIME'HIGH;
    else
        nextTime := EventCurrentTable(nextEventIndex).T + currentTime;
        nextEvent:= EventCurrentTable(nextEventIndex).T;
    end if;
nextEventIndex := NextEventIndex +1;
Result <= temp;
    end if;
end LOOP;
end;
```

2   EXAMPLE OF CURRENT MODELING

As an example of current modeling, the model of current flow at a two-input AND gate is proposed. For this, it will be necessary to define a set of parameters Ci (slope i),Pi (initial point I) and ti (time between two changes of the current flow). These parameters are defined by the VitalCurrentTableType type generic port. Each row is associated to a transition of the VitalTruthTableType type generic port. In this way, when an input of the AND gate changes from 0 to 1(symbol /), the output being 0, the form of the current will be given by the following pairs:

$$[0,0 \text{ A}],[500\text{ps}, 10 \text{ mA}], [1000 \text{ ps}, -1\text{mA}], [1200 \text{ ps}, 0 \text{ A}]$$

```
entity and2 is
    generic( InputTransition : VitalStateTableType( 0 to 3, 0 to 2):=( ( '/', '1', '0'),        -- 1
                                                                        ( '1', '/', '0'),        -- 2
                                                                        ( 'B', 'B', 'r'),        -- 3
                                                                        ( 'B', 'B', 'f'));       -- 4
            CurrentWaveform                       : VitalCurrentTableType( 0 to 3, 0 to 3):=(
    ( ((2.0e7,0.0), 500 ps), ((-2.2e7,10.0e-3), 500 ps ), ((5.0e6,-1.0e-3), 200 ps), ((0.0,20.0e-9),
    TIME'HIGH)),                                                                                 -- 1
    ( ((2.0e7,0.0), 500 ps), ((-2.2e7,10.0e-3), 500 ps ), ((5.0e6,-1.0e-3), 200 ps), ((0.0,20.0e-9),
    TIME'HIGH)),                                                                                 -- 2
    ( ((1.0e6,0.0), 1 ns), ((0.0,1.0e-3),TIME'HIGH),((0.0,0.0),0 ns),((0.0,0.0),0 ns)),         -- 3
    ( ((1.0e6,0.0), 1 ns), ((0.0,1.0e-3),TIME'HIGH),((0.0,0.0),0 ns),((0.0,0.0),0 ns)) )         -- 4
                );
            port( InA,Inb : in Std_logic;
            Y : InOut Std_logic;
            vdd: Out Dcurrent := (0.0,0.0));
    end ;
```

3   CONCLUSIONS

In this paper a current model has been presented for VHDL structural descriptions which follow the rules laid down by the VITAL standard. For this, it has been necessary to define types and subprograms which model the current flow in the cell. These elements have not been introduced in the VITAL standard.

On the other hand, the simulator results seem to show little discrepancy compared to those obtained with electrical simulators like SPICE[1]. This permits the current modeling and the application of IDDQ[4] and IDDT[5] to test circuits (main application of the proposed techniques). As a consequence of the application of the models presented here to fault simulation, better results have been achieved than those obtained by some non-VHDL commercial logic simulators [6], since:

1.- It enables the separate parts of the circuit under test to be modeled, since as many Dcurrent type signal can be used as there are separate power-supply parts of the circuit.

2.- It enables the maximum value of the static and dynamic current in the fault free circuit to be estimated, thus facilitating the design of the current sensor. In fact, the sensor could be incorporated as a Vital Level 0 cell in the system description.

3.- The current modeling is tool independent because the current behavior is defined in the VITAL library. Thus, the modeling is very flexible and portable.

At the moment effort is being made in two lines of development. On one hand, to improve the modeling in the case where a signal changes while the port is still affected by a previous change, and on the other hand, to study the propagation of X-values through the circuit.

4   REFERENCES

[1] Rouatbi, B. Haroun, A. Al-khalili. Power Estimation for Sub-Micron CMOS VLSI circuits. ICCAD92. 1992.

[2] H. Kriplani, F. Najm and I. Hajj. Maximum current estimation in CMOS circuits. 29th Design Automation Conference. 1992.

[3] T. Krodel. PowerPlay-Fast Dynamic Power Estimation Based on Logic Simulation. ICCD91. 1991.

[4] R. Aitken. A Comparation of Defect Models for Fault Location with Iddq Measurements. ITC93. 1993.

[5] J. Arguelles et al. Iddt Testing of Continuous-Time Filters. VLST95. 1995.

[6] SystemHilo 4.5. Release Notes. VEDA Design Automation. December 1994.

[7] Standard VITAL ASIC Modeling Specification. IEEE 1076.4. July 1995.

3

EDGAR: A PLATFORM FOR HARDWARE/SOFTWARE CODESIGN

A. J. Esteves, J. M. Fernandes, A. J. Proença

Departamento de Informática, Escola de Engenharia
Universidade do Minho
4709 Braga codex, Portugal

ABSTRACT

Codesign is a unified methodology to develop complex systems with hardware and software components. EDgAR, a platform for hardware/software codesign is described, which is intended to prototype complex digital systems. It employs programmable logic devices (MACHs and FPGAs) and a transputer-based parallel architecture. This platform and its associated methodology reduce the systems production cost, decreasing the time for the design and the test of the prototypes. The EDgAR supporting tools are introduced, which were conceived to specify systems at a high-level of abstraction, with a standard language and to allow a high degree of automation on the synthesis process. This platform was used to emulate an integrated circuit for image processing purposes.

1 INTRODUCTION

All the platforms used in codesign are not universal, in the sense that not all the systems can be implemented in a straightforward way. Additionally, those platforms are generally too expensive, since they have a large number of hardware resources. If these resources are not completely used for a significant number of systems, the ratio performance/cost is extremely low.

The EDgAR (*Emulador Digital Altamente Reprogramável*) platform was designed to achieve a high performance/cost ratio and to implement complex systems with critical time constraints, used in real-time applications (especially computer vision systems). However, the platform design was not significantly constrained by the particular aspects of these systems.

EDgAR is a FPGA-based platform that includes a transputer that can be linked to a parallel architecture. With the EDgAR platform, prototypes of complex digital systems can be obtained in a short period of time.

The recent development on the area of re-programmable components (FPLDs - Field Programmable Logic Devices) made them attractive to fast and efficiently create prototypes, because their complexity can achieve tens of thousands of equivalent logic gates, and the manufacturers provide electronic CAD tools to support those components. Since the time of design and the production cost were reduced, and the FPLDs need no longer to be removed for programming, they can be used with success in codesign platforms.

The transputer is a microprocessor with communication and processing power and a simple interface. It allows the scale of parallelism, due to its capacity to be interconnected with other identical microprocessors.

Codesign is closely related to the design of systems with unreachable performance in software implementations, and systems with higher complexity than those implemented in hardware (ASICs) [1, 2].

This article is organised as follows. In section 2, the architecture of the EDgAR platform is described. The synthesis of digital systems with EDgAR is analysed in section 3, with comments to the different phases of the process: the system specification, the hardware/software partitioning, the allocation of platform resources to partitions, and the validation of the prototype obtained. In section 4, the emulation of a VLSI circuit, the GLiTCH, on EDgAR is presented.

2   THE ARCHITECTURE OF THE EDGAR PLATFORM

The structure of the EDgAR platform (figure 1) is supported by two major blocks:

  i) a digital information processing unit (UPDI), that implements a parallel computation node, with communication and scalar processing power, and where the digital signals processing speed is not crucial;

ii) a programmable logic unit (ULP), containing a great amount of reconfigurable resources and whose operation speed is close to that of the circuits with fast technologies available on the market, allowing better performances than those obtained with traditional simulators.

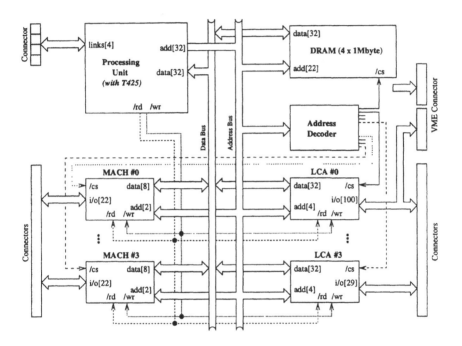

Figure 1 The architecture of the EDgAR platform.

To carry out the UPDI, the transputer (a microprocessor with communication
and processing power) was selected. It allows the scale of parallelism, due to its
capacity to be interconnected with other identical microprocessors, building up
a network with a variable topology. This processor is also implements the interface with the prototype development system and for the initial configuration
of the ULP components [3]. On the debugging phase, the user's interface with
the platform was developed on a unit containing several TRAMs (TRAnsputer
Modules) installed on a PC and using a C compiler. The connection between
the unit of TRAMs and EDgAR is done by one (or more) transputer link(s),
which are asynchronous. The tools available to work with the TRAMs allow to
monitor the transputers of the TRAMs and EDgAR, to compile the programs
and to load them to the transputers.

The ULP provides a large quantity of resources, without significantly compromising the speed of the systems being implemented. The ULP structure is based on two types of PLDs: one appropriated to implement circuits containing logic at two levels (MACHs - Macro Array CMOS High-density), while the other contains a structure organised like a matrix, suitable to implement circuits containing multi-level logic (FPGAs - Field Programmable Gate Arrays).

The present EDgAR platform version (figure 1) is implemented with a T425 transputer (a T805 could also be used), 4 Mbytes of DRAM, 4 MACHs and 4 FPGAs. The MACHs belong to the 2x0 AMD family, containing 44 pins, 64 macrocells and 32 I/O cells. The FPGAs are Xilinx LCAs that belong to the 3090A family: two FPGAs have 84 pins and the others have 175 pins. All FPGAs have 320 macrocells and 139 I/O cells.

All components are connected to common buses, using different addresses for the transputer internal and external memories, and for each of the FPGAs and MACHs. To emulate distinct digital systems on the platform, and to keep the possibility of reconfiguration by software, each MACH is connected to the buses by 2 address lines and 8 data lines, while each LCA uses 4 lines to connect to the address bus and 32 lines to the data bus. The remaining I/O pins of the MACHs and LCAs are available in connectors, allowing to emulate systems with different number of I/O signals and different size of hardware components. To scale the processing power, the transputer communication lines (links) are available outside the board. To scale the hardware resources, the VME connector can be used to link the FPGAs on EDgAR with other platforms that also have a VME bus.

3   DIGITAL SYSTEMS SYNTHESIS WITH EDGAR

The development process with the present platform runs through several phases from the specification to the implementation, going through the simulation and test (figure 2). Next, it is explained how these phases are being incorporated on the development environment that will support EDgAR.

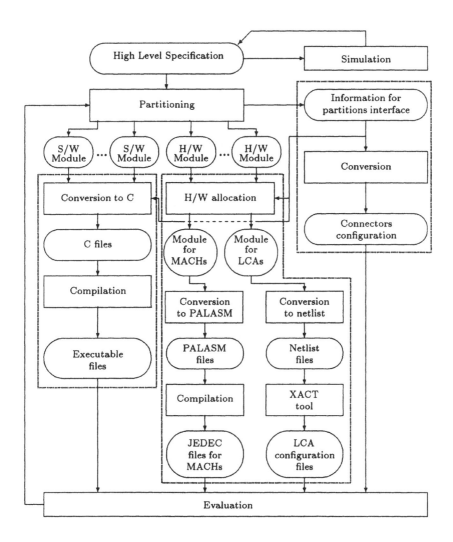

Figure 2 Methodology used for system development on the EDgAR platform.

3.1   Specification

On the codesign context, the selection of a high-level environment for system specification is being considered, which will be the basis of the specification model to be followed. The alternatives under study include an FSM-based representation, the OCCAM language, a representation using Petri Nets (PNs) or the VHDL language. A high level formal representation is used to prove the specification correctness and to guarantee that this correctness is preserved in the next design phases.

The modelling of systems with FSMs has two disadvantages: (i) as a high-level notation, FSMs are not so abstract as desired, and (ii) FSMs are not appropriate to represent systems with high algorithmic complexity [4].

The OCCAM language presents the advantages of being simple, suitable for real-time representation, having potential for parallelism, a well defined semantics (based on CSP [5]) and the adequacy to represent components to be implemented on the transputer [1]. OCCAM is not a good solution, because it is not a widely used language (this is reflected in the reduced number of available synthesis tools) and it has a strong binding to the transputer processors' family, which means that it is not an implementation independent language.

PNs are a mathematical formalism used to model systems that include concurrent activities and its graphical representation can be used to animate the modelled systems. The formalism associated with PNs allows the systems validation in relation to a set of properties: determinism, deadlock freedom, conflict freedom, liveness and boundedness [6].

VHDL is a standard hardware description language used to design digital systems, allowing the model to be clearly specified, simulated and synthesised. The specifications of the systems designed with VHDL can be hierarchically structured and properly represented [7].

The joining between VHDL and PNs is considered to be an acceptable solution. This was studied and applied with success in the specification of parallel controllers [8]. An identical evaluation is being carried out on the EDgAR platform, to implement systems that are more complex than those already tested.

The specification model is influenced by the fact that the EDgAR platform implements systems asynchronously, since a completely synchronous specification model is less suitable to represent the aspects related to implementations in hardware and software, which are asynchronous by nature. Although an independent implementation specification is a goal, this is not commonly achieved.

3.2 Hardware/Software Partitioning

The hardware/software partitioning, considered to be the most complex phase on the codesign context, is a hard task to be fully accomplished by an automatic process. Usually the partitioning algorithm is fed with inputs (supplied by the designer) to assist the process. The partitioning task comprises the phases of assignment and scheduling, although some approaches use assignment only [9, 10].

The partitioning applied in EDgAR is behavioural, since it is done on the system specification. The behavioural partitioning has several advantages over the structural partitioning, but the most relevant is the fact that the impact of changes on the system's specification is smaller on the first one [11].

The approach used for partitioning belongs to the software-oriented solutions. This means that the starting point is a complete software implementation, and afterwards parts of the system are moved to hardware, based on time criteria.

The software and hardware partitions are intended to have different granularities: task level on software partitions and block level on hardware partitions. Hardware partitions are implemented with the ULP in EDgAR and the software partitions with the UPDI. Among the hardware partitions, those implemented with MACHs must be distinguished from those implemented with FPGAs.

The partitioning comprises the isolation of the parts with critical time constraints, which will result on hardware partitions; the remaining parts may result on software partitions. The definition and implementation of the communication strategies and interface between partitions is an important aspect to be considered on the partitioning phase. On EDgAR, the interface between two software partitions is implemented with memory positions and transputer channels. Virtual channels are used if the partitions are on the same processor, while physical channels are used if the partitions are on different processors. The interface between two hardware partitions uses registers and connectors, and the interface between a hardware and a software partition is implemented with the resources used in the two previously mentioned types of interface.

3.3   Synthesis of Components

The synthesis of components is divided in three main parts: the synthesis
of software partitions (left block of figure 2), the synthesis of hardware partitions (central block) and the synthesis of the interface between partitions (right
block). Each part can be seen as an allocation of resources that results on a
configuration.

The allocation of UPDI resources to software partitions is accomplished in two
phases. In the first, the high-level specification of these partitions is converted
into modules on an intermediate language (C). This task requires the existence
of a converter to C language, and the generated C modules are compiled to the
transputer machine code.

The allocation of ULP resources to hardware partitions results in allocating to
these partitions resources available in two types of PLDs: MACHs and FPGAs.
The decision about which type of PLD to allocate to each module is based on
the need of storage elements and the existence of critical time constraints.
Partitions that need a number of storage elements higher than a critical value
are allocated to FPGAs, while partitions that require a response faster than
a critical value are allocated to MACHs. If both conditions arise in the same
partition and it can not be partitioned again, several components are allocated
to this partition.

To configure the MACH devices, the compilation and the later mapping of
their resources are completed with the agreement of the hardware allocation.
The result is a JEDEC file for each allocated device. The hardware allocated
to the FPGAs determines their configuration. The first step to obtain this
configuration is to create an intermediate format file (netlist) that will be used
as input to the Xilinx Automatic CAE Tools (XACT). These tools generate the
binary configuration file for each allocated FPGA, defining the device operation,
after the mapping, placing and routing of the specification.

When the system is powered on, the transputers download the configuration
files to the FPGAs and establish their operation. Among the available ways
to send the configuration file to the FPGA, the peripheral mode was selected,
which sends the configuration on a byte basis. After the start-up, the FPGA
can be reprogrammed without a physical reset of the system.

3.4   Components Verification

XACT allows for two types of simulation, in order to verify the parts of the system implemented with FPGAs: functional and timing simulations. The functional simulation detects logical errors, while the time simulation tests the functionality under different conditions, like a higher temperature, a lower power or a slower process.

The obtained prototype can be validated at a higher level of abstraction in a process called co-simulation. The co-simulation is a time consuming task that demands a huge computation power. For these reasons, it was intended to use a simulation model adapted to parallel architectures [12]. This advantage results because the co-simulation process runs on part of the same architecture that is used to implement the simulated prototypes.

4   THE EMULATION OF A VLSI CIRCUIT WITH EDGAR

The emulation of the GLiTCH chip [13], an associative processor array designed for a VLSI circuit to apply on image processing, was used as a case study, to validate the physical structure of the EDgAR platform and to explore the capabilities of the platform for codesign (figure 3).

The GLiTCH is structured on 5 blocks: an array of 64 1-bit processing elements (PEs), each one with 68 bits of associative memory (CAM), a pattern router (PBL), a video shift register (VSR) with 64x8 bits, and an instruction decoder [14].

The specification of this case study was not carried out at a high-level of abstraction: the modules to be implemented with the hardware components (MACHs and FPGAs) were specified using VIEWlogic schematics, while those to be implemented in software (transputer) were specified in C. To specify PLDs, using the ViewPLD tool from VIEWlogic, the JEDEC format and, textual descriptions in ABEL or VHDL could also be used.

Although manually done, the partitioning process used the performance of the system as the main *criterium* for partition definition, but it also used the particular characteristics of each block. Using a large granularity (block level), two candidates emerged to be implemented in hardware: the CAM and the

VSR. Since the VSR operates in two directions (columns rotation and rows shift), one of these operations would have a low performance if implemented in software. This leads to implementing the VSR in hardware. As a first approach, the CAM did not result on a hardware partition, due to its large dimensions (64x68 bits), but the software implementation did not significantly degrade the overall performance of the system. Further hardware partitions were not created as the PBL and the PEs are strongly tied to the CAM. Since the CAM resulted on a software partition, these two blocks are implemented in software too, reducing the communication cost between both partitions.

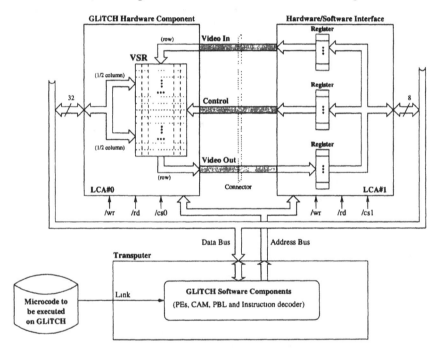

Figure 3   Hardware/software implementation of the GLiTCH on the EDgAR platform.

The VSR is a bi-dimensional shift register organised as a matrix. The GLiTCH uses an 8-bit video bus and includes 64 PEs, resulting on a VSR with 64x8 bits. The VSR functionality is represented by the operations performed on the data it stores. These operations are called SHIFT and SWAP, and correspond to row shift and column rotation, respectively. The SHIFT operation is regulated by the frequency of an external clock. This operation registers the 8 bits of the video input on VSR's row 63, it shifts all rows one position down, and row 0 is sent to the video output. The SWAP operation handles 64-bit columns, but the present implementation of this operation is done in two steps, because the data bus that connects the LCAs with the transputer is 32-bit wide. The SWAP operation reads column 0 to the data bus (parallel read), it registers the content of data bus on column 7, and it simultaneously rotates all the columns one position to the right (parallel write/column rotate). The SWAP operation is used to implement some GLiTCH instructions: rotate_image, extract_image and all others that use IMAGE as a parameter.

The hardware components of the GLiTCH emulator (VSR) was implemented in a 175-pin LCA. Two issues made the VSR implementation difficult: (i) the large percentage of the available storage elements allocated to the VSR (8*64=512), and (ii) the constraints imposed by the fixed position, on the PCB, of some signals (data, address and control). These two aspects result in problems: incomplete automatic routing of the LCA, long accumulated delays and fan-out problems. Some of these problems should be reduced, or even eliminated, if the VSR is implemented with 2 LCAs. However, this option would increase the cost associated with communication between the two VSR halves, and the chosen approach has the advantage of testing the utilisation of the LCAs on the limits (more than 80% of logic used).

To implement the software components of the GLiTCH emulator (PEs, CAM, PBL and instructions decoder blocks), the starting point was their functionality. The functionality of these blocks was described in ANSI C, but the emulator has some minor aspects especially developed for transputers [15]. The software components, running on a single transputer, fully implement the GLiTCH microinstructions, except those microinstructions using the VSR. If better performance is required, the parallel architecture connected to the platform should be used. Each microinstruction has one sub-operation executed by the PBL and one sub-operation executed by the PEs. The PBL sub-operation is executed before the PEs sub-operation (except in microinstructions that write to the CAM).

The interface between the hardware and the software components was implemented with 3 types of EDgAR resources: an 175-pin LCA, the data/address buses and the connectors. The FPGA is used to implement the VSR SHIFT operation, which is not synchronised by the same clock as the other GLiTCH components. The connectors establish the communication between the FPGA used in interface and the FPGA that implements the hardware partition.

The input to the GLiTCH emulator is the microcode of the several microinstructions to execute. For better interface with user, an assembler was developed.

5   CONCLUSIONS AND FUTURE WORK

The GLiTCH emulation led to the conclusion that the performance of the implemented systems strongly depends on the ULP resources allocated. The performance also depends on the hardware/software partitioning procedure. It is not expected that the level of abstraction used to specify the systems will significantly influence the final performance. The case study also demonstrates that EDgAR implements complex systems without scaling the platform, using connections to other platforms or computing nodes. The platform architecture was simplified because the transputer requires a simple interface and it supports the debugging of the architecture where it is included.

With the emulation of the GLiTCH processor using hardware and software components, significant improvements were obtained on the execution time of the instructions that use the VSR. Since the design time was not increased in the same proportion, it is demonstrated that the platform can be used successfully for hardware/software codesign.

The case study results in a hardware implementation without using any MACH, because the MACHs are devoted to implement fast combinational logic blocks, which are not present in the VSR. The validation of the MACHs was verified through other smaller sized systems.

When identical modules were implemented with both types of FPLDs, the delays achieved with FPGAs were bigger than the delays obtained with MACHs. This guarantees that, when both types of FPLDs are included on the platform, better performance is possible, since each device type is adequate to implement distinct parts of the system. This idea is represented by the two criteria used on the hardware partitions generation.

After the promising results obtained with EDgAR, the future work will be directed towards the integration on a more ambitious platform, which will include copies of an updated version of EDgAR, a microprogrammable unit based on a 16-bit sequencer and the MIMD transputer-based architecture. The VHDL language will be used as the unified specification notation, to improve the communication between the different phases of the codesign process: hardware/software partitioning, parallel co-simulation and synthesis.

While several tools for automatic synthesis are available, there is much work to be done for automatic partitioning and co-simulation. Future work includes: (i) the definition of a more complete partitioning strategy that automatically generates representations of the modules being implemented in FPLDs, the microprogrammable unit or the different transputer of the parallel architecture, and (ii) the development of a co-simulator that runs on the parallel architecture, whose main goal is to speed up the simulation, a generally time-consuming process.

6 REFERENCES

[1] Mike Spivey and Ian Page. *How to Design Hardware with Handel*, Oxford University Computing Laboratory, December 1993.

[2] Rajesh K. Gupta and Giovanni De Micheli. *System-level Synthesis using Re-programmable Components*. In *Proceedings of the European Conference on Design Automation*, pages 2–7, Brussels, Belgium, February 1992.

[3] António Joaquim Esteves. Rapid Prototyping of Digital Systems. Technical report, Dep. Informática, Universidade do Minho, Braga, Portugal, July 1994.

[4] M. Chiodo, P. Giusto, H. Hsieh, A. Jurecska, L. Lavagno, and A. Sangiovanni-Vincentelli. *A Formal Specification Model for Hardware/Software Codesign*. Technical report ERL-93-48, University of California - Berkeley, June 1993.

[5] C. A. R. Hoare. *Communicating Sequential Processes*. Prentice-Hall International, 1985.

[6] Manuel Silva and Robert Valette. Petri Nets and Flexible Manufacturing. In G. Rozenberg, editor, *Advances in Petri Nets 89*, volume 424 of *Lecture Notes in Computer Science*, pages 376–417. Springer-Verlag, Berlin, Germany, 1990.

[7] Douglas L. Perry. *VHDL*. McGraw-Hill, 1991.

[8] João Miguel Fernandes. Petri Nets and VHDL on the Specification of Parallel Controllers. Master's thesis, Dep. Informática, Universidade do Minho, Braga, Portugal, July 1994.

[9] Rolf Ernst, Jorg Henkel, and Thomas Benner. *Hardware-Software Cosynthesis for Microcontrollers*. *IEEE Design & Test of Computers*, December 1993.

[10] Asawaree Kalavade and Edward Lee. *A Global Criticality/Local Phase Driven Algorithm for the Hardware/Software Partitioning Problem*. In *Proceedings of the 3rd International Workshop on Hardware/Software Codesign*, pages 42–48, Grenoble, France. IEEE Computer Society Press, September 1994.

[11] Frank Vahid. *A Survey of Behavioral-Level Partitioning Systems*. Technical report 91-71, Dept. of Information and Computer Science, University of California, Irvine, October 1991.

[12] W. Billowitch. Simulation Models for Support Hardware/Software Integration. *Computer Design*, 1988.

[13] Henrique D. Santos, José C. Ramalho, João M. Fernandes, and Alberto J. Proença. A heterogeneous computer vision architecture: implementation issues. *Computing System in Enginneering*, 6(4/5):401–8, 1995.

[14] A. W. G. Duller, R. H. Storer, A. R. Thomson, E. L. Dagless, M. R. Pout, and A. P. Marriot. Design of an Associative Processor Array. *IEE Proceedings*, 136, 1989.

[15] António Esteves. Emulation of an Associative Processor Array with EDgAR Platform. Technical report UMDITR9602, Dep. Informática, Universidade do Minho, Braga, Portugal, May 1996.

4

HIERARCHICAL MULTI-VIEWS MODELING CONCEPTS FOR DISCRETE EVENT SYSTEMS SIMULATION

A. Aiello, J-F. Santucci, P. Bisgambiglia, M. Delhom

SDEM - URA CNRS 2053 University of Corsica
Quartier Grossetti B.P. 52
20250 CORTE (France)
e-mail: aiello,santucci,bisgambi,delhom@univ-corse.fr

ABSTRACT

In this article, we describe a Hierarchical Multi-Views Modeling Concepts for Discrete Event Systems Simulation. We give the basic concepts needed for the definition of the formal model allowing discrete events simulation of complex systems [1].

1 INTRODUCTION

This paper deals with basic concepts for modeling and simulation of complex systems. Four main features are needed to reach this goal:

- definition of different levels of abstraction - allowing to take into account the complexity of a system in a gradual way,

- definition of different levels of temporal granularity [2] - allowing to take into account only the pertinente informations at a given temporal level,

- definition of different views of a system - allowing to consider a system according to a structural or a behavioral view,

- definition of a generic simulation approach allowing to:

  - process the simulation of systems whatever the consideral view, level of abstraction or level of granularity,

– simulate different kinds of systems.

In order to define such an environment, we propose an original approach based on the simulation concepts introduced by B.P. ZEIGLER [3,4,5,6,7,8,9] and the modeling aspects developed in [10,11]. This approach has been used in order to study the behavior of complex systems such as a cachment basin [12]. We are currently applying the concepts introduced in this article for the design of embedded systems involving hardware and software components. In the second section, basic modeling concepts are presented. The third section is devoted to the description of the hierarchical multi-views modeling scheme. In the last part, we briefly present an object oriented simulation architecture.

2 BASIC MODELING NOTIONS

The modeling relation (see Fig. 1) allows to give simplified representation of system; the model is thus an image of the system. The model allows to study and anticipate the reactions of the system.

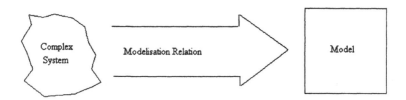

Figure 1   Modelisation relation

Four main concepts are used in our approach:

- multi-view notion - a system is represented by a set of distinct views corresponding to its differents aspects,

- abstraction hierarchy - a system can be described using different levels of details called abstraction levels,

- time hierarchy - a system can be considered under several temporal detail levels called "granularity levels",

- description hierarchy - a system can be described by a set of sub-systems at different description levels but at the same abstraction.

In order to take into account the different aspects of a model (structural, behavioral, time...), we propose two representations:

- the global representation: composed by all the desciptions of the model (the different views and the different levels of hierarchy). This hierarchical approach is made by different levels:

    - the abstraction levels
    - the time levels
    - the description levels

- the modeling space: is another expressive representation which consists in pointing out three dimensions:

    - the views
    - the abstraction levels
    - the time levels

- All the points of this space represent the model for:

    - a given view
    - a given abstraction level
    - a given time level

3 MODELING

This section deals with our modeling approach. Sub-section 3.1 gives the informal modeling scheme. Sub-section 3.2 gives the formal definition of the basic features of our approach.

3.1 Informal Model

We present how the system is represented in the three-dimension space (abstraction levels, time levels, views)[12]. We'll describe a set of functions which allows the translation between the different representations of the model.

The Structural View

The Structural View is composed by different abstraction levels which can be described under different time levels. A Structural View represents a potential structure of the system. It's an interconnexion of nodes. The nodes receive and send information by means of its input and output ports.

The Behavioral View

The Behavioral view doesn't represent a structure of the system but only its behavior. We have different levels of description. The behavior is expressed by means of a graph structure. In order to represent the behavior of one component, we use the "desc" function which provides the set of components which will allow to describe the behavior (description hierarchy). It is important to point out that a Structural View and a Behavioral View expressed by graph structure involve different concepts. The specification of the interconnexion of basic components represents the definition of a Coupled Model and the specification of a basic component represents the definition of an Atomic Model. A Coupled Model can be used like a basic component in a larger coupled model. Atomic and Coupled models have been defined in [6]. We present in the following a brief description of these models:

- A Coupled Model contains the following information:
    - the inputs
    - the ouputs
    - the names of the model components
    - the external input coupling
    - the external output coupling
    - the internal coupling
    - the priority list
- An Atomic Model provides a local description of a system's dynamic (figure 2).

• An Atomic Model provides a local description of a system's dynamic :

$$M = < X, S, Y, \delta int, \delta ext, l, ta >$$

X	: set of external input events
S	: set of sequential states
Y	: set of external output events
δint	: internal transition function
δext	: external transition function
l	: output function
ta	: time advance function

Figure 2

Translation functions

A set of functions have been defined in order to allow the translation of information between different levels and views:

- the *trans* function allows to transmit information between two abstraction levels.

- the *comp* function can be used for two kinds of operations: to decompose a model at a given abstraction level into a set of sub- models at another abstraction level.

- the *conv* function allows the transfer of information between the time levels.

- the *desc* function provides the set of components which compose a given model.

3.2 Formal Modeling Scheme

We give in this sub-section a formal expression of the basic features of our modeling scheme. Let us call MG the general model of a given system. MG is composed by the set of models M describing the system according to behavioral and structural views (Mc and Ms). Chv and Synth are two classes of functions allowing the translation of information between different views (see figure 3).

$$M_G = \{\ M, Niv_A, Niv_G, Chv, Synth\ \}$$

$$M = M_C \cup M_S$$

with M_C = behavioral model

and M_S = structural model

Niv_A = Set of abstraction level

Niv_G = Set of granularity level

$Chv = \text{map}\ (Niv_A, Niv_G, N_S) \rightarrow (Niv_A, Niv_G, Niv_D, N_C)$

$$(niv_a^i, niv_g^m, n_m^i) \mapsto (niv_a^i, niv_g^m, niv_d^{i}, {}^1 n_m^i)$$

$Synth = \text{map}\ (Niv_A, Niv_G, Niv_D, P(N_C)) \rightarrow (Niv_A, Niv_G, P(N_S))$

$$(niv_a^i, niv_g^m, niv_d^{q}, {}^q el_m^i) \mapsto (niv_a^i, niv_g^m, e2_m^i)$$

with $el \in P(N_C)$

and $e2 \in P(N_S)$

$P(N_C)$: set of the parts of N_C

$P(N_S)$: set of the parts of N_S

Figure 3

Formal Structural Model

We give in figures 4 and 5 a formal expression of the structural modeling scheme.

Formal Behavioral Model

In figures 6 and 7 we give a formal description of the behavioral modeling scheme. This description involves the definition of two sets of attributes: AMC and AMA (respectively the set of attributes of a Coupled Model and AMA the set of attributes of an Atomic Model).

$$M_S = \{ \ N_S, C_S, P_S, T_S^P, \psi, \varphi, \chi, \gamma, \text{Trans}_P, \text{Comp}_N, \text{Conv}_P \ \}$$

N_S = Set of the nodes of the structural model

C_S = Set of the node connexions of the structural model

P_S = Set of the ports of the structural model ($P_S = P_S^E \cup P_S^S$)

T_S^P = Set of the port types of the structural model

ψ = map $P_S \rightarrow N_S$

$\qquad p \mapsto n$

φ = map $N_S \rightarrow P(P_S)$

$\qquad n \mapsto p$

Figure 4

4 CURRENT AND FUTURE WORK

Our current work concerns the different aspects of the simulation and the implementation. Indeed, the simulation is divided in two parts:

- the structural part
- the behavioral part

We present the implementation of the simulation, and how the Object Oriented Programing's concepts [14,15] are used in order to obtain an evolutive, modular and hierarchical simulator. The architecture of the behavioral part of our environment is represented in Fig. 8.

We are currently developing the specifications of the general environment, with emphasis on the structural part of the simulation and the integration of the already implemented part. This environment will be used for the design and the validation of an embedded system in the framework of an international project [16]. This system will be a test bed for conducting trade-off studies between technologies, fault tolerance and testing in space. The designed system should

$$\chi = \text{map } C_S \rightarrow (P_S^S \cdot P(P_S^E))$$
$$c \mapsto (p_1, p_2)$$
$$\gamma = \text{map } P_S \rightarrow T_S^{P}$$
$$p \mapsto t$$
$$\text{with } p \in P_S = P_S^E \cup P_S^S$$
$$t \in \{in, out\}$$
$$\text{if } \quad \gamma(t) = in \quad \text{then} \quad p \in P_S^E$$

$$Trans_P = \text{map } (Niv_A, Niv_G, P(P_S)) \rightarrow (Niv_A, Niv_G, P(P_S))$$
$$(niv_a^i, niv_g^m, p_m^l) \mapsto (niv_a^j, niv_g^m, p_m^j)$$

$$Comp_N = \text{map } (Niv_A, Niv_G, N_S) \rightarrow (Niv_A, Niv_G, P(N_S))$$
$$(niv_a^i, niv_g^m, n_m^l) \mapsto (niv_a^i, niv_g^m, e_m^j)$$
$$\text{with } \quad e_m^j \in P(N_S)$$

$$Conv_P = \text{map } (Niv_A, Niv_G, P(P_S)) \rightarrow (Niv_A, Niv_G, P(P_S))$$
$$(niv_a^i, niv_g^m, p_m^l) \mapsto (niv_a^i, niv_g^m, p_n^l)$$

Figure 5

be able to provide data which characterize patterns of test defects occurring in space.

5 REFERENCES

[1] Y.C. Ho, "Dynamics of discrete event systems", Proceeding of the IEEE, Special issue on dynamics of discrete event systems, Vol.77, N 1, 1989, pp. 3-6.

$$M_C = \{\ N_C, P_C, AMC, AMA, Niv_D, Trans_P, Conv_P, Desc_{MC}, Desc_{MA}\ \}$$

N_C = Set of the nodes of the behavioral model

P_C = Set of the ports of the structural model

$AMC = \{\ I, O, C, COE, CIE, CI, LP\ \}$

    I = Set of the input ports of the coupled models and $I \subset P(P_C)$

    O = Set of the output ports of the coupled models and $O \subset P(P_C)$

    C = Set of the componants of the coupled models and $C \subset P(N_C)$

    COE = Set of the externe output couplings of the coupled models ·

    CIE = Set of the external input couplings of the coupled models

    CI = Set of the internal couplings of the coupled models

    LP = Set of the priority lists of the coupled models

Figure 6

[2] J. Euzenat, "Granularité dans les représentations spatio-temporelles", IN-RIA, rapport de recherche n 2242, Avril 1994.

[3] B.P. Zeigler, Theory of Modelling and Simulation, Wiley, New-York, 1976.

[4] B.P. Zeigler, Multifacetted Modelling and Discrete Event Simulation, Academic Press, 1984.

[5] B.P. Zeigler, "DEVS Representation of Dynamical Systems: Event-Based Intelligent Control", Proceeding of the IEEE, Special issue on dynamics of discrete event systems, Vol.77, N 1, 1989, pp. 72-80.

[6] B.P. Zeigler, Object-Oriented Simulation with Hierarchical, Modular Models, Intelligent Agents and Endomorphic Systems, Academic Press, 1990.

[7] T.I. Vren and B.P. Zeigler, "Concepts for Advanced Simulation Methodologies", Simulation J., Vol. 32, N 3, 1979, pp. 69-82.

[8] T.G. Kim and B.P. Zeigler, "The DEVS - Scheme Simulation and Modelling Environnement", Knowledge Based Simulation: Methodologies and Applications, Springer Verlag, New York, 1990, pp. 20-35.

$AMA = \{ \ X, Y, S, \Delta ext, \Delta \text{int}, \Lambda, TA \ \}$

  $X =$ Set of the input ports of the atomic models

  $Y =$ Set of the output ports of the atomic models

  $S =$ Set of the states of the atomic models

  $\Delta ext =$ Set of the external transition functions of the atomic models

  $\Delta \text{int} =$ Set of the internal transition functions of the atomic models

  $\Lambda =$ Set of the output functions of the atomic models

  $TA =$ Set of the time‑advanced functions of the atomic models

$Niv_D =$ Set of the description levels of the behavioral model

$Trans_P = \text{map} \, (Niv_A, Niv_G, Niv_D, P(P_C)) \rightarrow (Niv_A, Niv_G, Niv_D, P(P_C))$
$$(niv_a^i, niv_g^m, niv_d^q, {}^q p_m^i) \mapsto (niv_a^j, niv_g^m, niv_d^q, {}^q p_m^i)$$

$Conv_P = \text{map} \, (Niv_A, Niv_G, Niv_D, P(P_C)) \rightarrow (Niv_A, Niv_G, Niv_D, P(P_C))$
$$(niv_a^i, niv_g^m, niv_d^q, {}^q p_m^i) \mapsto (niv_a^i, niv_g^m, niv_d^q, {}^q p_n^i)$$

$Desc_{MC} = \text{map} \, (Niv_A, Niv_G, Niv_D, MC) \rightarrow (Niv_A, Niv_G, Niv_D, AMC)$
$(niv_a^i, niv_g^m, niv_d^q, {}^q mc_m^i) \mapsto$
$$(niv_a^i, niv_g^m, niv_{d, mc}^r {}^r i_{m, mc}^i {}^r o_{m, mc}^i {}^r c_{m, mc}^i coe_{m, mc}^i {}^r cie_{m, mc}^i {}^r ci_{m, mc}^i {}^r lp_m^i)$$

$Desc_{MA} = \text{map} \, (Niv_A, Niv_G, Niv_D, MA) \rightarrow (Niv_A, Niv_G, Niv_D, AMA)$
$(niv_a^i, niv_g^m, niv_d^q, {}^q ma_m^i) \mapsto$
$$(niv_a^i, niv_g^m, niv_{d, ma}^q {}^q x_{m, ma}^i {}^q y_{m, ma}^i {}^q s_{m, ma}^i {}^q \delta ext_{m, ma}^i {}^q \delta \text{int}_{m, ma}^i {}^q \lambda_{m, ma}^i {}^q ta_m^i)$$

Figure 7

[9] T.G Kim and S.B. Park, "The DEVS formalism: Hierarchical Modular Systems Specification in C++", Proceeding of the 1992 European Simulation Multiconference organized by the SCS, Modelling and Simulation, York, United Kingdom, 1992, pp. 152-156.

[10] C. Oussalah, "Modules hiérarchisés/multi-vues pour le support de raisonnement dans les domaines techniques", Doc. Université Aix-Marseille III, 1988.

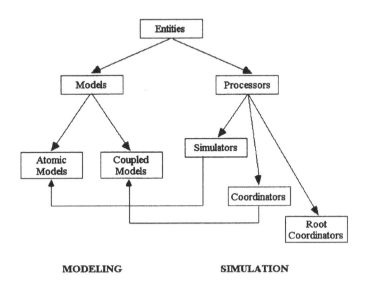

Figure 8 Architecture of the behavioral part

[11] W. Delaney and E. Vaccari, Dynamic Model and Discrete Event Simulation, Marcel Dekker, NY, 1989.

[12] M. Delhom, P. Bisgambiglia, J.F. Santucci, A. Aiello, "Modeling and Simulation of Discrete Event Systems: the study of the hydrologic behavior of a catchement basin",2nd IEEE International Conference on Systems, Man and Cybernetics, October 95, Vancouver, 22-25 October 1995, p.4190-4195.

[13] A. Aiello, J.F. Santucci, P. Bisgambiglia, M. Delhom, "A hierarchical Multi-Views Modeling and Simulation Environment for Discrets Event Systems", 3rd BELSIGN Workshop Proceding, 11-12 April 1996, Purticciu, Corsica, France.

[14] B. Stroustrup, "Le langage C++", InterEditions, 1989.

[15] B. Meyer, Object Oriented Software Construction, Prentice Hall, 1988.

[16] A.Rucinski, S.K. Tewksbury, B Dziurla-Rucinska, R. Harisson, P. Bisgam-biglia, J. F. Santucci, IEEE 2nd Workshop on "Hierarchical Test Generation", Sept 1995, p 40-41.

This work is supported by the EEC: HCM BELSIGN Network Contract n CHRX-CT 94-0459

A FORMAL SYSTEM FOR CORRECT HARDWARE DESIGN

M. Allemand, S. Coupet–Grimal, J–L. Paillet

Laboratoire d'Informatique de Marseille – URA CNRS 1787
39, rue F. Joliot–Curie 13453 Marseille France
e-mail: {amichel,solange,jluc}@gyptis.univ-mrs.fr

ABSTRACT

We present a formal system interpreted in a functional algebra. Well-formed expressions depict circuit architectures and their interpretations correspond to the associated behaviours. We also describe our approach for interfacing it with proof assistants (Coq and LP) in view of formal verification and synthesis. Then, we briefly comment our experiences in the field of formal proof.

1   INTRODUCTION

VLSI chips reliability is a critical question within the more general framework of the production of automated systems. Moreover, the manufacturing process cost makes it essential to detect design errors before the physical realization. Correctness verification is usually done by simulation. However exhaustive simulations required for zero defect design are untractable in case of large devices. Thus, the present trends are to use formal methods for mathematical validation of circuits.

Formal verification of digital circuits requires to abstract devices into mathematical objects, to carry out algebraic transformations on them, and finally to prove properties. The work presented in this paper is part of a more general study concerning a CAD verification oriented system for synchronous circuits (FORMATH: FORmal Modelling And THeorem provers), which satisfies these requirements.

Roughly speaking, it is composed of three parts:

- a formal system, called the P-calculus, which is the core of FORMATH.

- theorem provers, presently Coq and the Larch Prover (LP), and the interface between the formal system and the provers.

- user interfaces, for inputting specifications and descriptions of circuits: either by directly entering P-calculus expressions, or by translating current HDL (such as VHDL) descriptions into P-calculus.

Our purpose in this article is to develop the two first points above. The P-calculus is not a simple stream based functional HDL, but it is an actual formal system which allows algebraic transformations. Stream based functional modelling of hardware has been widely used and many functional HDL haven been already defined such as LCF-LSM [12], LUSTRE[14], muFP [22], HML [18] · · · (see also [15, 17] · · ·). These languages allow to describe and to simulate circuits, and sometimes to synthesize FSM, but they do not permit to directly process automated formal transformations.

In [19] a functional algebra was introduced which was named P-Calculus for the first time by Bronstein in [6]. We modify this initial algebra and enrich it with new operators. In addition we couple it with a formal system involving a typed formal language and a rewriting system which has been proven to be complete [2, 3]. The resulting new P-Calculus thus establishes a clear distinction between structures (described in a modular and hierarchical way by expressions of the formal system) and behaviours (functions of the algebra), both being linked by means of an interpretation function. Moreover, the rewrite rules correspond to semantics preserving transformations that are mechanically performed independently of the theorem prover to be used. Among other advantages, these preliminary transformations permit to detect early some kinds of errors and to simplify the proof process to come. Moreover, due to the readibility of the syntax, expressing specifications, transformations and refinements is straightforward.

The complexity of the mathematical demonstrations and of circuits makes it crucial to have proof processes validated by proof-assistants. Several demonstration tools are already well known and used in the community of the formal hardware verification. We can cite, for example: Nqthm [5], HOL [13], Nuprl [7], ... At present, we experiment with two provers LP [10] and Coq [9], which have very different features and thus are complementary in resolving the problems raised by such or such devices. An interface is devoted to the translation of the P-calculus expressions into the provers syntax. This interface also allows to carry out proof pre-processing independently of the provers. After doing some transformations such as normalizations, after detecting errors and possibly reducing the complexity of the problem to be resolved, the interface provides uniform expressions that are translated in a precise prover syntax.

The first section presents the main features of the P-calculus : the formal system with its language, its interpretation in the functional algebra, and a decomposition result obtained by rewrite-rules. Then we give in a second part a brief description of Coq and LP as well as the essential aspects of the interface between the P-calculus and these provers. This section ends by a short discussion about the experience we got from various examples that we have handled. More details about the proofs can be found in [2, 1, 8].

2 THE P–CALCULUS

2.1 The functional algebra

Let us start by giving some basic definitions. We first introduce the notion of *temporal sequences* that models signals and of *sequential functions* that are functions on these sequences. We shall deal only with sequential circuits synchronized by one clock whose cycles are formally represented by the naturals. Thus a signal x in a circuit will be a sequence $(x(t))_{t \in I\!N}$, so called a *temporal sequence*, whose set of values is a *basic type* (such as, in practice, booleans, naturals $\cdots$). The sequence $x = (x(t))_{t \in I\!N}$ represents the history of the wire x. In the following, the word *sequence* will mean *temporal sequence* and $\mathcal{S}_T$ will denote the set of all the temporal sequences of type T.

The behaviour of a circuit can be modelled by a function, called a *sequential function*, which associates a sequence vector (the output vector) with a sequence vector (the input vector).

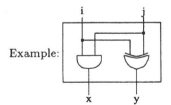

Example:

i, j, x, y are boolean sequences. The behaviour of this half-adder is a sequential function:

$$h\_add : \mathcal{S}_{I\!B} \times \mathcal{S}_{I\!B} \quad \to \quad \mathcal{S}_{I\!B} \times \mathcal{S}_{I\!B}$$
$$(i, j) \quad \mapsto \quad h\_add(i, j) = (x, y)$$

The basic combinational components (without temporal dependency) are concrete realizations of arithmetical and boolean functions. From these functions, defined on the basic types, *sequential functions* can be defined on the sequences. For example, *and* induces the *sequential function* $\overline{and}$: $S_B \times S_B \to S_B$ defined by:

$$\overline{and}(i,j) = (and(i(t), j(t)))_{t \in N}$$

Thus, the function h_add, which models the behaviour of a half-adder can be expressed in different ways. Starting from the expression of h_add on the input (i,j) at time t:

$$h\_add(i,j)(t) = (xor(i(t), j(t)), and(i(t), j(t)))$$

its abstraction is: $h\_add(i,j) = (\overline{xor}(i,j), \overline{and}(i,j))$ that we write, in a more concise way

$$h\_add = [\overline{xor},\ \overline{and}] = \overline{[xor,\ and]}$$

A sequential function F, which is obtained in such a way from a function f (one writes $F = \overline{f}$) is called a *projective* function. Note that not all the sequential functions are projective. More precisely the following result is proved in [2].

Proposition *A sequential function F is projective if and only if, for all sequence vectors X and Y*

$$\forall t, t' \in N \quad X(t) = Y(t') \implies F(X)(t) = F(Y)(t')$$

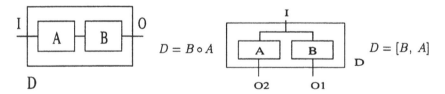

Figure 1 Composition   **Figure 2** Construction

The *composition* of sequential functions expresses the connection in series of two modules (fig. 1) and the *construction* represents the composition in parallel of two modules with the same inputs (fig. 2). Finally, the selection operators Sel_i ($i \in N^*$) denote the ith projections.

Example: see figure 3

It is easy to prove that the class of the projective sequential functions is stable for the *composition* and the *construction*, and that it contains all selectors.

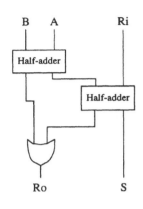

The behaviour of the adder on the left can be expressed by the equation:

$$ADD(Ri, A, B) =$$
$$(Sel_1(h\_add(Ri, Sel_1(h\_add(A, B)))),$$
$$\overline{or}(Sel_2(h\_add(Ri, Sel_1(h\_add(A, B))))),$$
$$Sel_1(h\_add(A, B)))$$

or, in a pure functional way without variables (close to that in [4]):

$$ADD =$$
$$[Sel_1 \circ h\_add \circ [Sel_1, Sel_1 \circ h\_add \circ [Sel_1, Sel_3]],$$
$$\overline{or} \circ [Sel_2 \circ h\_add \circ [Sel_1,$$
$$Sel_1 \circ h\_add \circ [Sel_1, Sel_3]],$$
$$Sel_1, Sel_1 \circ h\_add \circ [Sel_1, Sel_3]]]$$

Figure 3 An adder

When extending our purpose to the sequential circuits, it becomes necessary to take into account temporal constraints, and to include a new operator: the *past operator* $\mathcal{P}$. This operator is defined on every sequence x by:

$$\forall t \in I\!N \quad \mathcal{P}(x)(t+1) = x(t).$$

and in fact, it models a register:

$$\boxed{\begin{array}{c} x \\ \text{REG} \\ Px \end{array}}$$

The definition of $\mathcal{P}$ is extended to sequence vectors by:

$$\mathcal{P}(x_1, \cdots, x_n) = (\mathcal{P}(x_1), \cdots, \mathcal{P}(x_n))$$

Note that $\mathcal{P}(x)$ has an undefined value for $t = 0$. This corresponds to the fact that the initial value of a register is not part of the circuit architecture.

The past operator $\mathcal{P}$ is obviously not projective. A useful property of the past operator $\mathcal{P}$ is that it commutes with any projective sequential function. That formally expresses classical retiming.

Example:

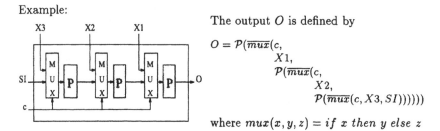

The output O is defined by

$$O = \mathcal{P}(\overline{mux}(c,$$
$$X1,$$
$$\mathcal{P}(\overline{mux}(c,$$
$$X2,$$
$$\mathcal{P}(\overline{mux}(c, X3, SI))))))$$

where $mux(x, y, z) = if\ x\ then\ y\ else\ z$

Figure 4   A serial shift register with parallel inputs

2.2   The formal system

We define a typed language in which the well-formed formulae will describe circuit architecture. Then each expression is interpreted by a function that represents the semantics (the behaviour) of the circuit.

We first introduce the simple P-calculus, in which circuits without loop will be described. Then we enrich it with a recursion operator for handling general devices.

Simple P-calculus

Syntax
The set of *object types* is defined as the least set which contains types of sequences on basic types and which is closed for the cartesian product.
In addition, from any two object types T_1 and T_2 and by means of the constructor "$\rightarrow$", new types $T_1 \rightarrow T_2$ can be defined. They are called *functional types*.

Vocabulary

- With each functional type T we associate:

    - a set V_T of functional variables of type T (which will be interpreted as projective sequential functions)

  – a set C_T of functional constants of type T (interpreted as sequential
  function corresponding to constant functions)

- with each object type T_o is associated a family $(P^n)_{n>0}$ of type $T_o \to T_o$
  (P^n will represent the nth power of $\mathcal{P}$)

- for all $m \geq 1$, for all object types $T_1, \cdots, T_m$ and for all $i \in \{1, \cdots, m\}$
  an element S_i of type $T_1 \times \cdots \times T_m \to T_i$ is given (in the interpretation it
  will be the ith projection).

The vocabulary consists of all the elements defined above and of four additional symbols $\odot$ (later interpreted as the functional composition), "," (coma), "[" and "]" (which will be used for representing the construction operator).

Expressions
The expressions of the simple P-calculus (P-expressions) are defined as follows.
Let T, T', T'' be three object types:

- for all $c \in C_{T \to T'}$, c is an P-expression of type $T \to T'$

- for all $m \geq 1$, for all object types $T_1, \cdots, T_m$ and for all $i \in \{1, \cdots, m\}$,
  S_i is a P-expression of type $T_1 \times \cdots \times T_m \to T_i$

- if e is a P-expression of type $T \to T'$, if P^n is of type $T' \to T'$ and if
  $f \in V_{T' \to T''}$ then $P^n \odot e$ is a P-expression of type $T \to T'$ and $f \odot e$ is a
  P-expression of type $T \to T''$

- If e_1 and e_2 are P-expressions of type $T \to T'$ and $T' \to T''$ respectively,
  then $e_2 \odot e_1$ is a P-expression of type $T \to T''$

- Let $T_1, \cdots, T_n$ be object types, if $e_1, \cdots, e_n$ are P-expressions of type
  $T \to T_1, \cdots, T \to T_n$ respectively, then $[e_1, \cdots, e_n]$ is a P-expression of
  type $T \to T_1 \times \cdots \times T_n$

Nothing else is a P-expression.

Example:
Let us consider the shift register with parallel loading described in figure 4.
Let us take $(X1, X2, X3, SI, c)$ as input vector and let MUX be a functional

variable of type $S_{I\!B} \times S_{I\!B} \times S_{I\!B} \rightarrow S_{I\!B}$. The structure of the circuit is modelled by the following P-expression:

$$e_{shift\_reg} = P^1 \odot MUX \odot [S_5, S_1, P^1 \odot MUX \odot [S_5, S_2, P^1 \odot MUX \odot [S_5, S_3, S_4]]]$$

Interpretation

The interpretation of the P-expressions depends on an arbitrary interpretation of the functional variables and of the functional constants. Let I be an application which:

- with each functional variable $f \in V_{T \rightarrow T'}$ associates a projective sequential function $I(f) : T \rightarrow T'$

- with each functional constant $c \in C_{T \rightarrow T'}$ associates the sequential function corresponding to a constant function $I(c) : T \rightarrow T'$.

This application I extends to all the P-expressions in the following way.

Let $e, e_1, \cdots, e_n$ be P-expressions, let f be a functional variable:

- $I(S_i) = Sel_i$,

- $I(P^n \odot e) = \mathcal{P}^n \circ I(e)$,

- $I(f \odot e) = I(f) \circ I(e)$,

- $I([e_1 \cdots e_n]) = [I(e_1), \cdots, I(e_n)]$

- $I(e_1 \odot e_2) = I(e_1) \circ I(e_2)$

It is easily verified that the types of the expressions insure the consistency of the compositions and the constructions.

Example: on the device in figure 4, the output O is described by the equation:

$$\begin{aligned} O &= \mathcal{P}(\overline{mux}(c, X1, \mathcal{P}(\overline{mux}(c, X2, \mathcal{P}(\overline{mux}(c, X3, SI)))))) \\ &= I(e_{shift\_reg})(X1, X2, X3, SI, c) \end{aligned}$$

In fact, the interpretation I defines the semantics of the P-expressions and thus the functional behaviour of the circuits. Therefore two expressions can be said to be *equivalent* when the associated circuits have the same behaviour, that is when the expressions have the same interpretation.

Characterizing temporal and combinational parts

Let us present this notion on the example of the shift register in figure 4. As $\mathcal{P}$ commutes with any projective sequential function we transform the equation defining the output O in the following way:

$$
\begin{aligned}
O &= \mathcal{P}(\overline{mux}(c, X1, \mathcal{P}(\overline{mux}(c, X2, \overline{mux}(\mathcal{P}(c), \mathcal{P}(X3), \mathcal{P}(SI)))))) \\
&= \mathcal{P}(\overline{mux}(c, X1, \overline{mux}(\mathcal{P}(c), \mathcal{P}(X2), \overline{mux}(\mathcal{P}^2(c), \mathcal{P}^2(X3), \mathcal{P}^2(SI))))) \\
&= \overline{mux}(\mathcal{P}(c), \mathcal{P}(X1), \overline{mux}(\mathcal{P}^2(c), \mathcal{P}^2(X2), \overline{mux}(\mathcal{P}^3(c), \mathcal{P}^3(X3), \mathcal{P}^3(SI))))
\end{aligned}
$$

The last expression corresponds to the *normal form* of the expression of O.

Let us set:

$$
P(X1) = V1 \ ; \ P^2(X2) = V2 \ ; \ P^3(X3) = V3 \ ; \ P^3(SI) = V4 \qquad (5.1)
$$
$$
P(c) = V5 \ ; \ P^2(c) = V6 \ ; \ P^3(c) = V7
$$

The equation becomes:

$$
O = \overline{mux}(V5, V1, \overline{mux}(V6, V2, \overline{mux}(V7, V3, V4)))
$$

This last expression of O represents the combinational part of the circuit. On the other hand, all the temporal features are expressed by the equalities 5.1.

Such a decomposition can be viewed as the construction of the equivalent circuit in figure 5.

Formally, this decomposition method corresponds to a rewrite system on the P-expressions the completeness of which has been proven in [2]. This rewrite system automatically generates the normal form of an expression. This is of interest, among other things, for syntactically characterizing certain classes of circuits and for simplifying proof processes.

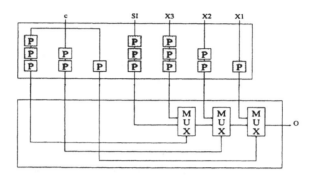

Figure 5   Shift register – Temporal and combinational parts

This transformation must not be confused with the classical decomposition method illustrated by the figure 6.

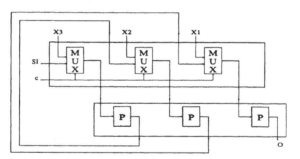

Figure 6   Shift register – Classical decomposition

Recursive P-calculus

In order to take into account structural loops, a recursion operator must be added to the P-calculus. Indeed, behaviours of circuits with loops cannot be defined algebraically. They are sequential functions obtained as the least fixed point, if any, of an equation. Our approach, close to [16] is explained in details in [2].

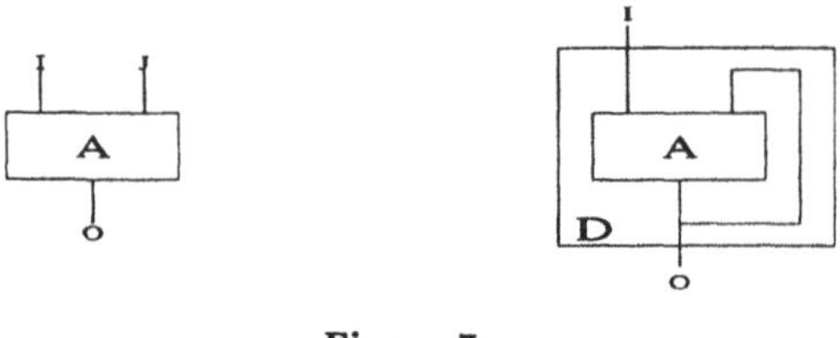

Figure 7

In such a way, we can model a module all outputs of which are connected to inputs (see figure 7). It can be shown that all other forms of recursions in circuit structures boil down to this particular case. Thus we only consider circuits D depicted in figure 7. Let n and m be the sizes of the vectors I and J. The informal idea is to describe the behaviour of the circuit D by a sequential function, still called D such that:

$$D(I) = A(I, D(I))$$

These considerations justify the following definition.

Definition *Let A : $\mathcal{S}^{n+m} \to \mathcal{S}^m$ be a sequential function. We define the sequential function $REC(A)$: $\mathcal{S}^n \to \mathcal{S}^m$ as the least solution (the less defined solution), if any, of the equation:*

$$REC(A) = A \circ [\mathcal{I}\lceil_{\mathcal{S}^n}, REC(A)]$$

We have proved that in case of well synchronized circuits in which every loop contains at least one register, and for each set of registers initial values, the equation has one and only one solution.

Thus, we are led to introduce an additional symbol R, in the simple P-calculus, which will be interpreted by the recursive operator REC.

Expressions of the recursive P-calculus
The expressions of the recursive P-calculus are defined as follows.

- Every expression of the simple P-calculus is an expression of the recursive P-calculus and its type is unchanged.

- Let e be an expression of the simple P-calculus of type $T_1 \times T_2 \to T_2$. Then $R(e)$ is an expression of the recursive P-calculus of type $T_1 \to T_2$

Finally, it remains to extend the definition of the interpretation I to these new expressions.

Interpretation
Let $e, e_1, \cdots, e_n$ be expressions of recursive P-calculus, let f be a functional variable then:

- $I(S_i) = Sel_i,$

- $I(P^n \odot e) = \mathcal{P}^n \circ I(e),$

- $I(f \odot e) = I(f) \circ I(e),$

- $I([e_1 \cdots e_n]) = [I(e_1), \cdots, I(e_n)]$

- $I(e_1 \odot e_2) = I(e_1) \circ I(e_2)$

Let e be an expression of the simple P-calculus.

- $I(R(e)) = REC(I(e))$

Example:

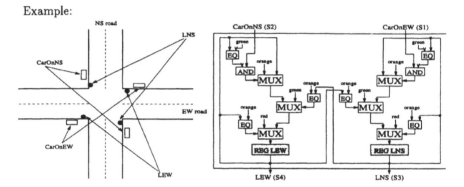

Figure 8   Road intersection with a traffic light and its implementation

Let us consider the traffic light controller (figure 8) proposed in [23]. The device detects by means of sensors (CarOnNS,CarOnEW) the presence of cars waiting on a North-South road (NS) and on an East-West road (EW). According to the french protocol and depending on the color of the lights, it makes the expected changes.

This circuit is described by the following P-expression where the inputs are two boolean signals (representing the sensors) and the outputs are the light colors.

$TRLIGHT =$
$\quad R([P \odot MUX \odot [EQC \odot [S_3, \overline{orange}],$
$\quad\quad\quad\quad \overline{red},$
$\quad\quad\quad\quad\quad MUX \odot [EQC \odot [S_4, \overline{orange}],$
$\quad\quad\quad\quad\quad\quad \overline{green},$

$$MUX \odot [AND \odot [S_1, EQC \odot [S_3, \overline{green}]], \overline{orange}, S_3]]],$$
$$P \odot MUX \odot [EQC \odot [S_4, \overline{orange}],$$
$$\overline{red},$$
$$MUX \odot [EQC \odot [S_3, \overline{orange}],$$
$$\overline{green},$$
$$MUX \odot [AND \odot [EQC \odot [S_4, \overline{green}], S_2], \overline{orange}, S_4]]]]$$

3   USE OF THEOREM PROVERS

3.1   Formath interface between the P-calculus and the provers

A P-calculus architectural description of a circuit consists of two parts:

■ a P-calculus expression which describes the structure of the circuit,

■ an interpretation of the functional variables occurring in the expression.

This amounts to consider the functional variables as black boxes only described by their behaviour. The user is given a lexicon of functional variable identifiers with predefined semantics. In each prover these identifiers will be the name of functions implementing the interpretation of the variables.

Moreover, starting from a P-calculus description, if the expression is of the form $R(A)$, FORMATH interface first generates the normal form (according to the transformations of the formal system) of the expression A. On the example of the traffic light controller, it results in:

$$TRLIGHT =$$
$$R([MUX \odot [EQC \odot [P \odot S_3, \overline{orange}],$$
$$\overline{red},$$
$$MUX \odot [EQC \odot [P \odot S_4, \overline{orange}],$$
$$\overline{green},$$
$$MUX \odot [AND \odot [P \odot S_1, EQC \odot [P \odot S_3, \overline{green}]],$$
$$\overline{orange},$$
$$P \odot S_3]]],$$
$$MUX \odot [EQC \odot [P \odot S_4, \overline{orange}],$$

$\overline{red}$,
$MUX \odot [EQC \odot [P \odot S_3, \overline{orange}],$
$\qquad \overline{green},$
$\qquad MUX \odot [AND \odot [EQC \odot [P \odot S_4, \overline{green}], P \odot S_2],$
$\qquad \quad \overline{orange},$
$\qquad \quad P \odot S_4]]]])$

Then the following points are automatically performed:

- A precedence graph is built. It describes the dependencies between signal values in the circuit at the same step of time.

- By means of a topological sort of this graph, the fact that each structural loop in the circuit includes at least one register is checked.

- In this case, recursive equations describing the outputs, are generated. Providing that initial values are given, these equations will be translated into recursive definitions in the syntax of the suitable prover.

Example:
Let us consider the informal example of a circuit whith input $I = [I_1, I_2]$ and output $O = [O_1, O_2, O_3]$, described by means of the recursive equation

$$O = F(I, O)$$

where $F = [F_1, F_2, F_3]$.
Assume that this equation is normalized (according to the transformations) in a system of the form:

$$\begin{aligned}
O_1 &= F_1'(P(O_1), P^2(O_1), P(O_2), O_3) \\
O_2 &= F_2'(O_1, P(O_2), P(O_3), P^4(O_3)) \\
O_3 &= F_3'(P^3(O_3))
\end{aligned} \qquad (5.2)$$

where no P occurs in F_i'.

By introducing the time variable t, the following system is automatically produced:

$$\begin{aligned}
O_1(t+2) &= F_1'(O_1(t+1), O_1(t), O_2(t+1), O_3(t+2)) \\
O_2(t+4) &= F_2'(O_1(t+4), O_2(t+3), O_3(t+3), O_3(t)) \qquad (5.3) \\
O_3(t+3) &= F_3'(O_3(t))
\end{aligned}$$

Moreover the initial values are required:

$$\begin{array}{llllll}
O_1(0) &= v_{(1,0)} & \qquad O_2(1) &= v_{(2,1)} & \qquad O_3(0) &= v_{(3,0)} \\
O_1(1) &= v_{(1,1)} & \qquad O_2(2) &= v_{(2,2)} & \qquad O_3(1) &= v_{(3,1)} \\
O_2(0) &= v_{(2,0)} & \qquad O_2(3) &= v_{(2,3)} & \qquad O_3(2) &= v_{(3,2)}
\end{array}$$

Starting from the system 5.3 the precedence graph (figure 9) expressing that O_2 depends on O_1 and O_1 depends on O_3 at the same step of time, is mechanically built.

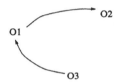

Figure 9   precedence graph of F

Then, a topological sort is performed in order to verify that the device is correctly synchronized (i.e. that the gragh is acyclic) and in order to produce a computation linear ordering of the outputs.

On our running example, it produces the following equations:

$TRLIGHT1(BS1, BS2)(t+1) =$
    $mux(eqc(TRLIGHT1(BS1, BS2)(t), orange),$
        $red,$
        $mux(eqc(TRLIGHT2(BS1, BS2)(t), orange),$
            $green,$
            $mux(andb(BS1(t), eqc(TRLIGHT1(BS1, BS2)(t), green)),$
                $orange,$
                $TRLIGHT1(BS1, BS2)(t))))$

$TRLIGHT2(BS1, BS2)(t+1) =$
    $mux(eqc(TRLIGHT2(BS1, BS2)(t), orange),$

$$red,$$
$$mux(eqc(TRLIGHT1(BS1, BS2)(t), orange),$$
$$green,$$
$$mux(andb(BS2(t), eqc(TRLIGHT2(BS1, BS2)(t), green)),$$
$$orange,$$
$$TRLIGHT2(BS1, BS2)(t))))$$

Here, *eqc* is an identifier interpreted by the equality on the set of colors.

Then it demands two initial values (one for each output) resulting in a primitive recursive definition.

3.2   The Larch Prover

The LP proof assistant is a rewrite rule based tool which works on a subset of typed first-order logic with equality. The basis for proofs in LP is a logical system called a "theory". Theories can be defined by means of sorts, variables, operators and properties on these operators. Sorts are sets of values. The properties on the basic objects are axiomatized by: equations which are automatically oriented into rewrite rules, operators theories (commutative or associative and commutative), deduction rules which are the basis for generating new equations from existing ones, induction rules defining a sort in terms of bottom and constructor functions.

A conjecture in LP is either a deduction rule or an induction rule or an equation of the form "$A == B$". Proving a conjecture consists in either rewriting it into true or proving some subconjectures the verification of which is sufficient to validate the initial one. Thus LP provides two inference mechanisms: backward and forward inference. The former produces consequences from a logical system; the latter yields a set of subgoals to be proved in order to validate a conjecture.

LP has not been widely experimented in the field of formal proof of hardware. The most significant previous studies include: the proof of circuits specified by means of synchronized transitions [11] and the proof of a simplified ALU by validation of some transformations [21]. These approaches are based on proof by cases and critical pairs computation but don't involve the LP proof by induction mechanism.

Implementing the P-calculus into LP

To each basic type in the P-calculus corresponds in LP the declaration of two sorts, one for the type and another one for the temporal sequences of this type.

Example:
For the color type, we declare the sorts *color* and *sequences of color*:
```
declare sort Color, Color_seq
```

The value of a sequence at time t is defined by means of an overloaded operator dot "." with the signature:
```
declare operator .  :  T_seq, Natural -> T
```

Thus, the sequences being considered as objects instead of functions, second order equations can be expressed although LP only supports first order. A wire of a circuit is implemented by a constant of a sequence sort. Sequence vectors are implemented by their components.

Example:
The wires $CarOnNS, CarOnEW, LEW$ and LNS of the traffic light controller are implemented by:

```
declare operators CarOnNS, CarOnEW : -> Bool_seq
declare operators LEW, LNS : -> Color_seq
```

A functional variable F of type $S_{T_1} \times \cdots \times S_{T_n} \to S_T$, which is interpreted by a sequential function $\overline{f} = I(F)$ is implemented by the equation:
`(F(X1, ··· , Xn)).t == f(X1.t, ··· , Xn.t)` where the Xi are sequence variables.

Example:
The identifier AND will be associated with the following declarations:

```
declare operators AND : Bool_seq, Bool_seq -> Bool_seq
declare variables X, Y : Bool_seq
assert AND(X,Y).t == (X.t) & (Y.t)
```

We need to make the distinction between the general description of the module *ADD* and the description of a particular instance of this module.

The *construction* of a t-uple of modules is simply described by the implementation of each modules. In the same way, the description of the composition of two modules is done by the implementation of the instance of each module where the outputs of one of them are the inputs of the other one.

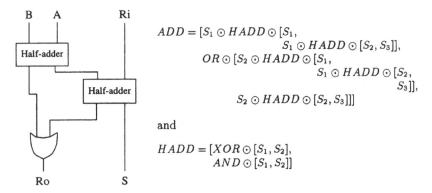

$$ADD = [S_1 \odot HADD \odot [S_1,$$
$$S_1 \odot HADD \odot [S_2, S_3]],$$
$$OR \odot [S_2 \odot HADD \odot [S_1,$$
$$S_1 \odot HADD \odot [S_2,$$
$$S_3]],$$
$$S_2 \odot HADD \odot [S_2, S_3]]]]$$

and

$$HADD = [XOR \odot [S_1, S_2],$$
$$AND \odot [S_1, S_2]]$$

Figure 10   The adder

Example:
The adder in figure 10 is described by the following declarations:

```
declare operaror XOR, AND, OR, HADD1, HADD2 : Bool_seq, Bool_seq -> Bool_s
declare operator ADD1, ADD2 : Bool_seq, Bool_seq, Bool_seq -> Bool_seq
declare var S1,S2,S3 : Bool_seq
assert
   HADD1(S1,S2) == XOR(S1,S2)
   HADD2(S1,S2) == AND(S1,S2)
   ADD1(S1,S2,S3) == HADD1(S1,HADD1(S2,S3))
   ADD2(S1,S2,S3) == OR(HADD2(S1,HADD1(S2,S3)),HADD2(S2,S3))
..
declare operator A,B,Ri,S,Ro : -> Bool_seq
assert
   S == ADD1(Ri,A,B)
   Ro == ADD2(Ri,A,B)
..
```

In case of circuits with loops the recursive definitions are generated by the interface (see 3.1).

Example:
This is the description of the first output of the traffic light controller:

```
declare op
    TRLIGHT1, TRLIGHT2 : Bool_seq, Bool_seq -> Color_seq
    CarOnNS, CarOnEW : -> Bool_seq
    LEW, LNS : -> Color_seq
..
declare variables BS1, BS2 : Bool_seq
  TRLIGHT1(BS1,BS2).(t+1) ==
        MUX(EQ(TRLIGHT1(BS1,BS2).t, orange_s.t),
            red_s.t,
            MUX(EQ(TRLIGHT2(BS1,BS2).t, orange_s.t),
                green_s.t,
                MUX(ET(BS1.t, EQ(TRLIGHT1(BS1,BS2).t, green_s.t)),
                    orange_s.t,
                    TRLIGHT1(BS1,BS2).t)))
  TRLIGHT2(BS1,BS2).(t+1) ==
        MUX(EQ(TRLIGHT2(BS1,BS2).t, orange_s.t),
            red_s.t,
            MUX(EQ(TRLIGHT1(BS1,BS2).t, orange_s.t),
                green_s.t,
                MUX(ET(BS2, EQ(TRLIGHT2(BS1,BS2).t, green_s.t)),
                    orange_s.t,
                    TRLIGHT2(BS1,BS2).t)))
  TRLIGHT1(BS1,BS2).0 == green
  TRLIGHT2(BS1,BS2).0 == red
assert
  LNS == TRLIGHT1(CarOnEW, CarOnNS)
  LEW == TRLIGHT2(CarOnEW, CarOnNS)
..
```

3.3   The Coq proof assistant

A higher order typed λ-Calculus

The Coq system is based on the calculus of inductive constructions [9]. This is a higher typed λ-Calculus in which definitions of mathematical objects can

be given and proofs of propositions on these objects can be performed. Both
objects and propositions are terms of the Lamba-Calculus.
Moreover, there are two kinds of types: the propositions are of sort Prop and
the sets are of sort Set.

These are the building rules for the terms:

- identifiers refer to defined constants or to variables declared in a part called
  context

- $(A\ B)$ denotes the application of functional object A to B

- $[x : A]B$ abstracts the variable x ot type A in term B in order to construct
  a functional object, that is generally written $\lambda x \in A.B$ in litterature

- $(x : A)B$ as a term of type Set corresponds to a product $\prod_{x \in A} B$ of a familly
  of sets B indexed on A. As a term of type Prop, it corresponds to $\forall x \in A\ B$.
  If x does not occur in B, $A \rightarrow B$ is a short notation which represents either
  the set of all the functions from A to B or a logical implication.

Induction principles

A typical example of inductive definition in Coq is:

$$Inductive\ Set\ nat = 0 : nat\ \mid S : nat \rightarrow nat$$

which defines the set of naturals as the smallest set containing 0 and its successors. Referring to this definition, the system provides two elimination principles:

- The non dependant elimination principle. A function $f : nat \rightarrow C$ is
  defined, according to a primitive recursive scheme, by means of a given
  element x of C and a function $H : nat \rightarrow C \rightarrow C$. f verifies:

$$
\begin{aligned}
(f\ 0) &= x \\
\forall n \in nat\ (f(Sn)) &= (H\ n\ (f\ n))
\end{aligned}
$$

Remark: In Coq f is written: `[n:nat] (<C> Match n with x H)`

- The dependant elimination principle which corresponds to the well known recurrence principle. Let $C : nat \rightarrow Prop$ be a predicate. To find a proof f of $\forall n \in \mathbb{N}\ C(n)$, that is to obtain a term $f : (n : nat)(C\ n)$, it is sufficient to build two terms:

$$x : (C\ 0)$$

$$H : (n : nat)(C\ n) \rightarrow (C(S\ n))$$

Enumerated types can be given by means of non-recursive inductive definitions. For example the type *bool* of booleans is defined by

```
Inductive Set bool= true:bool | false:bool
```

and the type *color* of the light colors is defined by

```
Inductive Set color = red:color | orange:color | green:color.
```

In these cases the elimination principles correspond to definitions or proofs by cases. For example, the *and* connector on booleans is defined by:

```
Definition andb = [b1,b2:bool](<bool> Match b1 with
                    b2          (* return b2 if b1 is true *)
                    false).     (* return false if b1 is false *)
```

and the equality on the set of colors is given by:

```
Definition eqc:color->color->bool= [c1,c2:color]
(<bool>Match c1 with
    (*red*)(<bool> Match c2 with true false false)
    (*orange*)(<bool> Match c2 with false true false)
    (*green*)(<bool> Match c2 with false false true)).
```

Extracting programs

In this system, proving a proposition P, amounts to construct a term of type P. Therefore, a proof is a λ-term and thus a program. One of the salient feature of Coq is the possibility of extracting an algorithm from a proof [20]. For example, an algorithm computing the natural D can be extacted from a proof of the proposition:

$$\exists D/(D \mid a) \wedge (D \mid b) \wedge (\forall d, (d \mid a) \wedge (d \mid b) \rightarrow d \leq D).$$

which specifies the GCD.

Some parts of the proof term can be considered as logical comments, the other parts being purely computational. Adequate declarations permit to make a synctatical distinction between logical and computational parts. The extraction mechanism consists in keeping only the computational parts, resulting in the algorithm correct by construction.

Implementation of the P-calculus into Coq

We take advantage of the higher order and define a generic sequence type, the parameter of which is a basic type.
$$\text{Definition Seq} = [\text{T:Set}] \text{ nat->T}.$$
Thus (Seq bool) is a type of boolean sequences.

The sequential function AND is expressed by:

```
Definition AND = [B1, B2:(Seq bool)][t:nat] (andb (B1 t) (B2 t)).
```

In this way, we can describe the half-adder and the adder in figure 3 by:

```
Definition Half_Adder=[I,J:(Seq bool)](Pair(XOR I J)(AND I J)).
Definition Adder =[Ri,A,B: (Seq bool)]
  (Pair
    (S1 (Half_Adder Ri (S1 (Half_Adder A B))))
    (OR
        (S2 (Half_Adder Ri (S1 (Half_Adder A B))))
        (S2 (Half_Adder A B)))).
```

As we did with LP, the implementation of the traffic light controller uses the outputs generated by the interface between the P-calculus and the provers. The circuit is described by the following declarations.

```
Definition Mux:=[b:bool][x,y:color](<color>Match b with(*true*)x(*false*)
Definition Output: nat->color*color= [t:nat]
(<color*color> Match t with
 (*0*)     (Pair red green)
 (*(S p)*)  [p:nat][Previous_Output:color*color] (Pair
    (Mux (eqc (s1 Previous_Output) orange)
          red
          (Mux (eqc (s2 Previous_Output) orange)
               green
               (Mux (andb (eqc (s1 Previous_Output) green) (CarNS p))
                    orange
                    (s1 Previous_Output)))
    (Mux (eqc (s2 Previous_Output) orange)
          red
          (Mux (eqc (s1 Previous_Output) orange)
               green
               (Mux (andb (eqc (s2 Previous_Output) green) (CarNS p))
                    orange
                    (s2 Previous_Output)))))).
Definition LEW:(Seq color)=[t:nat] (s1 (Output t)).
Definition LNS:(Seq color)=[t:nat] (s2 (Output t)).
```

3.4   General ideas on the verification part

Our purpose in this section is not to compare, as it is usually done in case of fully automated provers, proof CPU times. This does not make sense in this context, since proof process times are insignificant and moreover they are negligible with respect to the time required from the user for establishing mathematical strategies. In addition, a proof in Coq is a Lambda-term which is constructed step by step by means of tactics. At the end of the process, this term is saved and thus it is available, if needed, without demanding the proof to be run again.

We handled several examples of synchronous sequential circuits in both provers, investigating their potentials and taking advantage of their particular features in order to get proof processes as general, as easy to drive, as neat, and as understandable as possible. Among the devices we studied, three of them seem particularly relevant in support of our first conclusions.

In the case of control dominated circuits such as the traffic light controller presented in this paper, the superiority of LP is undeniable. This is due to the fact that in the Coq underlying intuitionnist logic, a proposition is not interpreted by a boolean value (as in the classical model theory). Proving a proposition does not consists in showing that the semantics value of the proposition is true: a proposition is a type and proving it amounts to finding out a lambda term inhabiting this type. Therefore, boolean operators must be defined by the user and the proof requires user intervention whereas in LP, as the boolean calculus is built-in, the proof is almost automatic.

We also verified a multiplier by iterated additions given in [12]. In this case, the proofs are almost similar although Coq requires more expertise from the user. However, we can take advantage of Coq higher order for proving universal theorems that can be applied for any proof by invariant, as it is the case in this example. This leads to more general and more complete proofs.

The third significant circuit we studied implements the well known algorithm of the so called egyptian multiplication. As in the case of the previous multiplier, the proof is classically performed in LP by finding out an invariant and by proving it. We chose a radically different approach with Coq, since we synthesized the circuit starting from its specification. One of the main points of the synthesis process relies on an extraction of the algorithm from a proof, as indicated in the presentation of Coq given in the previous paragraph. Among others advantages of this method, no invariant needs finding out.

4   CONCLUSION

In this paper we focused on the theoretical part of FORMATH. This verification CAD system relies on the P-calculus, a formal system interpreted in a functional algebra. Well-formed expressions depict circuit architectures and their interpretations correspond to the associated behaviours. The P-calculus has been completely described and grounded. Moreover, an interface with proof tools has been defined. It essentially compiles the P-expressions, detecting loops without registers, if any (in case of sequential circuits which are not well-synchronized) and removing all occurrences of the past operator P. The output is a set of equations depending on the time and it is straightforwardly translated into primitive recursive definitions in the syntax of such or such prover, provided a set of initial register values is given.

The main advantage of the P-calculus is that it provides an uniform framework for formally describing and specifying circuits, and for performing behaviour preserving algebraic transformations on them. Thus, it is a convenient tool in the perspective of the "design for verifiability" which aims at integrating verifications at each step of a design process, going from abstract specifications up to concrete implementations.

In addition of our current work in the field of the verification (including investigations on proof and synthesis strategies with respect of several theorem provers and of different kinds of circuits), our projects concerning the P-calculus are to develop a bi-directional translator between the P-calculus and VHDL to be integrated in an user interface. Due to VHDL widespread use in the international community, this point cannot be evaded and demands a specific investigation related to VHDL denotational semantics.

5 REFERENCES

[1] M. Allemand. Formal verification of characteristic properties of circuits with the larch prover. In R. Kumar and T. Kropf, editors, *Theorem Prover in Circuit Design: Theory, Practice and Experience*, Bad Herrenalb, Germany, Sept. 1994. FZI–Publication.

[2] M. Allemand. *Modélisation formelle et preuve de circuits avec LP*. PhD thesis, Université de Provence, July 1995.

[3] M. Allemand, S. Coupet-Grimal, and J.-L. Paillet. A functional algebra for circuit modelling and its implementation in LP. Research Report 1995.099, LIM, Mar. 1995.

[4] J. Backus. Can programming be liberated from the von neumann style? a functional style and its algebra of programs. *Communications of the A.C.M.*, 21(8):613 – 641, Aug. 1978.

[5] R. S. Boyer and J. S. Moore. *A computational logic*. ACM Monograph Series. Academic Press Inc., 1979.

[6] A. Bronstein and C. Talcott. Formal verification of synchronous circuits based on String-Functional Semantics: The seven Paillet circuits in Boyer-

Moore. In *Workshop on automatic verification methods for finite state systems*, Grenoble, June 1989.

[7] R. Constable, S. Allen, H. Bromley, W. Cleaveland, J. Cremer, R. Harper, D. Howe, T. Khnoblock, N. Mendler, P. Panangaden, J. Sasaki, and S. Smith. *Implementing Mathematics with Nuprl Proof Development System*. Prentice Hall, 1986.

[8] S. Coupet-Grimal and L. Jakubiec. Vérification formelle de circuits avec COQ. In *Journées du GDR Programmation*, Sept. 1994.

[9] G. Dowek, A. Felty, H. Herbelin, G. Huet, C. Murthy, C. Parent, C. Paulin-Mohring, and B. Werner. The Coq Proof assistant User's Guide - version 5.8. Technical Report 154, INRIA, Inria-Rocquencourt - CNRS - ENS Lyon, May 1993.

[10] S. J. Garland and J. V. Guttag. A guide to LP, the Larch Prover. Report 82, DEC Systems Research Center, Palo Alto, CA, Dec. 1991.

[11] S. J. Garland, J. V. Guttag, and J. A. Staunstrup. Verification of VLSI circuits using LP. In *The Fusion of Hardware Design and Verification*, pages 329–345, Glasgow, July 4–6 1988. IFIP WG 10.2, North Holland.

[12] M. Gordon. LCF-LSM. Technical Report 41, University of Cambridge, 1984.

[13] M. Gordon. HOL: A proof generating system for higher-order logic. In G. Birtwistle and P. Subrahmanyam, editors, *VLSI Specification, Verification and Synthesis,* pages 73–128. Kluwer Academic Publishers, 1988.

[14] N. Halbwachs, A. Lonchampt, and D. Pilaud. Describing and designing circuits by means of synchronous declarative language. In D. Borrione, editor, *IFIP WG 10.2 Workshop From HDL descriptions de guaranted correct circuit designs*. North-Holland, 1987.

[15] S. Johnson. *Synthesis of Digital Designs from Recursive Equations*. The MIT Press, Cambridge, 1984.

[16] G. Kahn. The semantics of a simple language for parallel programming. In *IFIP Congress, Information processing 74*. North-Holland, 1974.

[17] J. T. O'Donnell. Hardware description with recursion equations. In M. Barbacci and C. J. Koomen, editors, *Computer Hardware Description Languages and their applications*, Amsterdam, Apr. 1987. North Holland.

[18] J. O'Leary, M. Linderman, M. Leeser, and M. Aagard. HML: a hardware description language based on standard ML. Technical Report EE-CEG-92-7, Cornell School of Electrical Engineering, Oct. 1993.

[19] J.-l. Paillet. A functional model for descriptions and specifications of digital devices. In D. Borrione, editor, *IFIP WG 10.2 Workshop From HDL descriptions de guaranted correct circuit designs*. North-Holland, 1987.

[20] C. Paulin-Mohring. *Extraction de programmes dans Coq*. PhD thesis, Université Paris 7, 1989.

[21] J. B. Saxe, S. J. Garland, J. V. Guttag, and J. J. Horning. Using transformations and verification in circuit design. *Formal Methods in System Design*, 3(3):181 –209, Dec. 1993. Also published as DEC Systems Research Center Report 78 (1991).

[22] M. Sheeran. mufp, a language for VLSI design. In *ACM Symposium on Lisp and functional programming*, pages 104 –112, Austin, Texas, 1984.

[23] J. A. Staunstrup, S. J. Garland, and J. V. Guttag. Mechanized verification of circuit descriptions using the Larch Prover. In V. Stavridou, T. F. Melham, and R. T. Boute, editors, *Theorem Provers in Circuit Design: Theory, Practice, and Experience*, pages 277–299, Nijmegen, The Netherlands, June 1992. IFIP TC10/WG10.2, North-Holland, IFIP Transactions A-10.

6

INTEGRATION OF BEHAVIORAL TESTABILITY METRICS IN HIGH LEVEL SYNTHESIS

K. Olcoz*, J.F. Santucci**, J.F. Tirado*

*Depto. de Informática y Automática,
Univ. Complutense. 28040 Madrid (Spain)
**Faculté des Sciences et Techniques,
Univ. of Corsica. Quartier Grossetti, BP 52, 20250 Corte (France)
katzalin@dia.ucm.es, santucci@univ-corse.fr, ptirado@dia.ucm.es

ABSTRACT

Testability is an important aspect of digital systems usually addressed after the design phase. In this paper we propose to take into account testability constraints at the beginning of this phase. We therefore propose a novel approach for performing high level synthesis according to area and testability constraints. Behavioral testability metrics are defined and used during data path allocation. Furthermore, we will point out how testability and area considerations are merged in a high level synthesis system to obtain a global exploration scheme of the design space.

1 INTRODUCTION

This paper deals with a novel approach for the high level synthesis of digital systems according to area and testability constraints. Testing is one of the most important aspects when producing digital devices. In order to take into account testability features in a high-level synthesis process, we include Behavioral Testability Metrics [1][2] in a high level synthesis system called FIDIAS [3]. Testability considerations in FIDIAS are inserted during data path allocation, so metrics have to be defined from data available at that moment of the design cycle, namely: the scheduled control-data flow graph, SDFG, and the partial design, PD (part of the data path already allocated). For every allocation there are usually several options, marked by their increments on area and testability. Testability metrics are used with area estimations to select options with better testability area relationship. Rules for selection of allocation alternatives that increase testability are developed based on the metrics, to fasten the search in the design space. Finally, a meta-rule is provided to combine area saving rules with rules for controllability and observability increasing. It must select which kind of rules (based on area or testability considerations) is used in the first place. It is founded on the initial value of the testability-area priority given by the user. This initial value can be modified by the design expert if either the testability or the area constraint are not satisfied. As a consequence, design exploration is flexible and can obtain designs with different testability-area tradeoffs. The paper is organized as follows. In the first part, testability metrics are presented. The second part gives a brief overview of allocation in the FIDIAS system; area-test tradeoffs made during design space search are introduced in this part also. In the third part we present rules for testability maximization. First, the scheme for controllability and observability merging is detailed and then, the different rules are explained with the help of some examples.

2 COMPUTATION OF TESTABILITY METRICS

Three testability metrics are defined, listed in growing complexity order: distance based, functional and reconvergence-divergence. We propose several ways to define behavioral testability metrics to compensate the fact that they are not totally accurate. The computation of totally accurate metrics is known to be an NP-complete problem that would request too much computational effort. In order to calculate these three different metrics, we have defined two graph models representing the SDFG and the PD. Next, the models are defined in sub-sections 2.1 and 2.2 and then, computation of the metrics is outlined in sub-sections 2.3, 2.4 and 2.5.

2.1 Model of SDFG

The behavior to be allocated is shown as a control-data flow graph, made up by nodes and variables representing both control and data dependencies. Every node in the graph has been scheduled in one control step. From the testability metric point of view, components of the SDFG are divided into four classes:

- operational node (OPN): it represents a function performed on the input variable(s),

- assignment node (ASN): it is a data transfer,

- divergence: a divergence point is a variable used as input of several nodes and a divergence node is either the output of a variable in a loop or the input of a variable in a conditional branch,

- reconvergence node: it is the input of a variable in a loop or the output of a variable in a conditional branch.

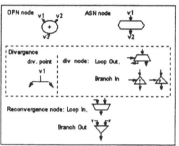
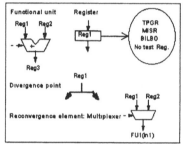

Figure 1: SDFG model Figure 2: DP model.

In figure 1, a representative of each class of element is displayed with its data (solid) and control (dashed lines) inputs and outputs.

2.2 Model of data path

Since the partial design is the portion of the data path that has been allocated to the moment, testability of the partial design has to be based on a model of the data path. It is a set of connected hardware elements taken from a library of modules. Regarding testability computation, they are classified as follows (see figure 2):

- functional unit (FU): element performing an operation,

- register: storage element. They can be further separated according to their test abilities (test generators, compactors, both or none),

- divergence point: it represents a connection feeding several inputs,

- reconvergence element: a multiplexer.

2.3   Distance-based metric

Testability of an element is defined in terms of its controllability and observability, that are computed as the closeness to controllable and observable elements respectively. So, controllability is the maximum closeness from any input element and observability the maximum closeness to any output element. They are measured in number of nodes and hardware elements for the SDFG and PD. The closeness between to elements E1 and E2 is derived from:

$$Closeness(E_1, E_2) = 1 - \frac{Distance(E_1, E_2)}{1 + Max.\, distance}$$

where Distance(E1,E2) is the minimum number of elements traversed to reach E2 from E1 and Max. distance is the maximum distance along the DFG or DP. The maximum distance has been defined for the DFG as the number of control steps minus one. This definition is based on the hypothesis that there is a path connecting an element in the first control step to one in the last control step and that such a path has a node in each control step. It is a reasonable choice because it is easy to compute and quite accurate since most schedules are minimal or nearly minimal (high level synthesis tools rarely increase the number of control steps and, when doing so, they only add 1 or 2 to the former time).

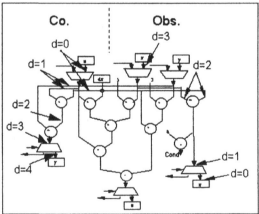

Figure 3: EcuDif. On the left, distances
from input elements are shown.
On the right, distances to output elements.

The maximum distance in the data path is computed alike: a non cyclic path from an element used in the first control step and another used in the last one is assumed and no wait control steps are allowed in it. But, this time the number of hardware elements traversed in a control step is variable from 1 (only a register) to 4 (a multiplexer at the input of a FU, the FU, a multiplexer at the input of a register and the register). Since the complete data path is not known until the end of the allocation process, the number of elements that make up each path cannot be known. So an additional assumption is needed: the maximum (4) is used as number of elements in one control step. Then, the maximum distance is computed as four times the number of control steps minus one. Distance computation for every element is done by the algorithms shown in figures 4 (distance to an output element needed for observability computation) and 5 (distance from an input element required to obtain controllability). An example of testability computation for a DFG commonly used in high level synthesis (the differential equation solver in [4]) is displayed in figure 3.

```
dist(Out) = 0;
From control step N to 1
      dist-in(OPN) = dist-out(OPN) +1 ∀J
      dist-in(ASN) = dist-out(ASN) +1
      dist-in(div. point) = min {dist-outJ(div. point)}
      dist-in(div. node) = min {dist-outJ(div. node)} +1 ⊥
      dist-inJ(rec. node) = dist-out(rec. node) +1 ∀J
Check on loops
```

Figure 4: Distance to Out computation.

2.4 Functional metric

Testability is no longer dependent on the number of elements but on their operation. Since functional metrics only take into account the functionality of elements traversed, testability is modified only for operational elements (OPN nodes or functional units). Thus, the controllability of the output of an element is the same as the one of the input except for operational elements. For them, the controllability is computed as the product of the controllability of the input by the control transfer factor of the operator (CTF). It represents the fraction

```
dist(In) = 0;
From control step 1 to N
    dist-out(OPN) = max {dist-inJ(OPN)} +1
    dist-out (ASN) = dist-in(ASN) +1
    dist-outJ(div. point) = dist-in(div. point) ∀J
    dist-outJ(div. node) = dist-in(div. node) +1 ∀J
    dist-out(rec. node) = min {dist-inJ(rec. node)} +1 ⊥
Check on loops
```

Figure 5: Distance from In computation.
⊥ Control lines are fully controllable.

of data that can be reached at the output for that operation (cardinal of the operator's image set divided by all possible values). Observability of input k is computed quite alike, save that not only the observability of the output is included but also the controllability of input $l \neq k$.

```
obs(Out) = 1;
From control step N to 1
    obs-inJ(OPN) = obs-out(OPN) * OTF(operator)
    * min {co-inK(OPN)} ∀J
    obs-in(div.) = max{obs-outJ(div. point)}
    obs-in(rest) = obs-out
OFT(operator) = fraction of non dominant values in the domain set
of the operator. ♣
```

Figure 6: Observability computation.

co(In) = 1;
From control step 1 to N
    co-out(OPN) = min{co-inJ(OPN)}*CTF(operator)
    co-out(rec. node) = max {co-inJ(rec. node)}
    co-out(rest) = co-in
CTF(operator) = fraction of all binary configurations in
the image set of the operator. ♣

Figure 7: Controllability comp. (F).
♣ They are computed by simulation.

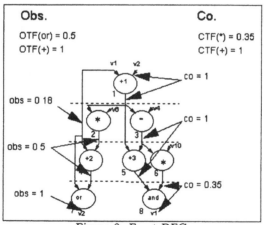

Figure 8: Facet DFG

Algorithms for observability and controllability computation are shown in figures 6 and 7 respectively. Their application to a high level synthesis example (the Facet DFG in [5]) can be seen in figure 8.

co(In) = 1;
From control step 1 to N
    co-out(OPN) = min {co-inJ(OPN)}
    * CTF(operator) * CD(OPNnode)
    co-out(rest) = like functional metric
CD(OPNnode) = fraction of output values reachable in
spite of data dep. between inputs.

Figure 9: Controllability computation (R-D).

2.5 Reconvergence-divergence metric

Finally, reconvergence-divergence metrics take into account the effect that data
dependencies, due to divergences that later reconvert, have on the controllability of different inputs of an operational element. Then, with this more precise
controllability figure (see figures 9 and 10 for details), controllability of the output of an operational element and observability of every input are computed as
for the functional metrics. Due to its big complexity, this metric is only applied
to the SDFG. The reason is that testability of the SDFG is computed only once
(before allocation) while testability of the PD must be estimated every time an
allocation choice is taken.

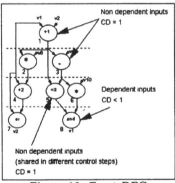

Figure 10: Facet DFG

3 SEARCH GUIDING IN ALLOCATION

Allocation in FIDIAS uses a branch and bound algorithm that allows exploration of different designs for the same behavior. The sequence in which these designs are found determines the search time. In order to reduce exploration time some heuristics are defined. Their goal is that the best data paths in terms of testability and area are obtained as soon as possible. The method we have defined sorts out the list of alternatives for each allocation (that is, the list of functional units for an operational node and the list of registers for storage of results) using the heuristics. The result is that alternatives with better testability and area are the first elements to be tried in.

Good designs correspond to designs having big testability and small area. They are reached when allocation decisions produce testability increments (gain) that are worth their counterpart area increment (cost).

For any allocation alternative, area increment is computed using the estimation explained in [6]. The testability increment caused by each alternative is obtained as follows: first, controllability increment for all the inputs and observability increment for the output are computed according to the metrics above introduced, and then, the maximum among the increments is selected as testability increment.

When design search is over, one data path is selected as best due to its area and test figures. Users should be allowed to establish the area-test priority for their designs. The best design according to the user criterion is the one with area-testability ratio that suits the user given priority. Thus, to obtain that design as quickly as possible, the order in which alternatives are explored should also be determined by the area-test priority (β). It represents the degree of test importance, so, when it takes a value close to 0, area is much more important than test, while figures close to 1 lead to test-driven searches.

To sum up, search time is saved if designs explored have good test-area relationship, so allocation decisions need to have good gain-cost ratio. Further, selection of the best design depends on the user criterion, and so does search guiding.

3.1 Global exploration scheme

Search guiding is performed by application of area and test rules, that sort out the list of available elements according to minimum cost and maximum gain respectively. Testability increment rules and the area saving rule have to be merged in a unique exploration heuristic. So, a meta-rule is needed to decide between area saving and testability increasing. This meta-rule fixes the search order, so that it determines how the design space is searched. The underlying idea is that cost (area) is spent if the gain (testability inc.) is worth it. The gain is worth if it is bigger than the opposite of the area-test priority so that the more important the area, the larger the gain required to spend it.

Let β be the area-test priority, and DTe[0,1] the testability increment (either in controllability or observability). If testability increment is bigger than $1 - \beta$, test rules are applied first. Else, area rule is applied first. So, if β is close to 0, the area rule would be applied in the first place unless testability increment was close to 1. On the other hand, bigger values of β cause selection of testability rules for smaller gains.

As a result of this scheme, the search should be speeded up because the searching criterion matches the quality criterion. Besides, design space exploration is flexible and can vary according to the user preferences. Finally, if the constraints are not met for a given value of the user criterion, the design expert can change that value to search for a valid design. This capability allows smart search of a wider design space.

4 TESTABILITY INCREASING RULES

They order allocation choices according to the testability increment implied. These rules are also twofold: controllability and observability rules. Controllability rules order the list of alternatives (functional units or registers) according to the controllability increment that would be caused by selection of the alternative. Observability rules work in the same way with regard to observability increment.

Controllability and observability rules have to be merged, that is, for a given allocation, the order in which both testability rules are tried in must be established. The merging scheme is to apply the rule that leads to maximum increment. If they produce the same gain, the controllability rule is preferred because observability depends on controllability and not the other way around.

4.1 Controllability rules

They are aimed at sorting out the list of available elements (FU or registers) according to their controllability increments. There are some different rules depending on the controllability of the connection's source(s): a controllable source improves the controllability of the element chosen as destination. So, the less controllable the destination element, the larger the controllability increment. An example is shown in figure 11, where a register must be selected to store data from an input port.

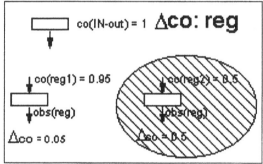

Figure 11: elements involved in reg.
selection for the output of IN.

On the other hand, if the source of data is not controllable, connecting it to an element will not improve the controllability of the source element. But, choosing a controllable element will improve the controllability of the variable. If this variable is later used as input, it will be more controllable and some controllability increment will be achieved. In figure 12 the source element is a non controllable FU output. If a controllable register is connected to it, the result generated by the FU becomes controllable.

To specify more precisely the rules, lets assume that an element is controllable if its controllability is bigger than β. Thus, the controllability value required for an element to be controllable is proportional to the test importance. **If** the source of data is controllable, the following rule is applied: controllability increment for each alternative is computed. **If** no controllability increment is big enough, it means that controllability of the source is wasted (all available elements are already controllable). So, creation of a new element is considered based on area increment, number of elements in the partial design and estimated minimum number of elements in the complete data path and on the impact of creating a controllable element (it is, whether or not the element will be further used).

Else, elements are selected to maximize controllability increment. Elements producing the same increment are sorted out following the observability rule. Rule for register selection is introduced as example in figure 13, where *out* means the output of the source element.

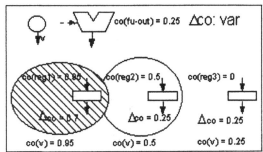

Figure 12: elements involved in register
selection for the output of FU.

Next, situation shown in figure 11 is used to illustrate the rule. Let $\beta = 0.4$ (area is a little more important than test), the order registers are tried in according to the controllability rule is: first, register 2, then register 1 and the third choice is creating a new register. If $\beta = 0.6$, the first choice would be creating a new register ($\Delta co = 1$) and only after, registers 2 and finally 1 would be tried in.

4.2   Observability rules

Their difference from controllability rules comes from the fact that observability is transmitted from the last to the first control steps, that is, opposite to allocation flow. This means that when an allocation has to be made, the

If co(out) $\geq \beta$
    for all available registers (regI)
        ΔcoI = co(out) - co (regI)
    if ($\Delta coI < \beta, \forall I$) & (new_reg = TRUE)
        create new register
    else
        select reg for maximum Δco
        if several, apply observability rule
    if there are no free register, create new

Figure 13: controllability rule for register selection
when the source is controllable.

controllability of the source element is known (paths from input ports to that element have already been allocated). On the contrary, observability of the destination element is only determined when a path to an output element is created, usually in the last control steps. Thus, not only observability of the output is considered but also observability of the variable in the SDFG (if it is big, a path to an output port will be created in the following control steps). So, according to the observability of the output variable (in the SDFG) there are different rules: if the variable is observable, it will improve the observability of the element chosen as destination. In such cases, the less observable the destination, the bigger the gain, as displayed in figure 14 for FU selection.

On the other hand, a non observable variable does not necessarily imply that the source of the connection is not observable. So, two more rules are needed for the two remaining possibilities. If both the source of the connection and the variable are not observable, an observable destination is searched to improve their observability (see figure 15).

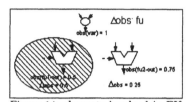

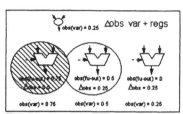

Figure 14: elements involved in FU selection when the variable is observable

Figure 15: elements involved in FU selection for a non observable node.

Finally, if the variable is not observable but the source is, no observability gain is got by selection of an observable destination, so the less observable ones are chosen. The rule applied for observable variable is: observability increment is computed. If no increment is big enough and trying to avoid observability waste, creation of a new element is considered. Else, the list of elements is arranged to maximize observability increment. Equal increments are sorted out by using the controllability rule. Application of this rule to figure 14, results in selection of FU 1 in the first place.

5   CONCLUSIONS AND FUTURE WORK

A scheme for the automatic synthesis of data paths has been outlined. It is based on testability metrics that are defined in a consistent way on the circuit's structure (data path) and behaviour (SDFG), so that testability of the unallocated part of the circuit is obtained from the second.

Integration of these metrics in the allocation step of a high level synthesis tool has also been proposed, in such a way that it drives exploration of the design space. Area and test guiding make exploration shorter and adaptable to different user priorities.

6   REFERENCES

[1] J.F. Santucci, G. Dray, N. Giambiasi, M. Boumedine, "Methodology to reduce computational cost of behavioral test pattern generation using testability measures", 29th IEEE/ACM Design Automation Conference, 1992, pp.267-272.

[2] J.F. Santucci, G. Dray, M. Boumedine, N. Giambiasi, "Methods to Measure and to Enhance the Testability of Behavioral Descriptions of Digital Circuits", 1st IEEE Asian Test Symposium, 1992, pp.118-123.

[3] J. Septièn, D. Mozos, J.F. Tirado, R. Hermida, M. Fernández, H. Mecha "FIDIAS: An integral approach to high-level synthesis", IEE Proc. on Circuits, Devices and Systems, vol. 142, no. 4, pp. 227-235, August 1995.

[4] P.G. Paulin, J.P. Knight, E.F. Girczyc, "HAL: A Multi-Paradigm Approach to Automatic Data Path Synthesis", Proc. of the Design Automation Conference, 1986, pp. 263-270.

[5] Ch. Tseng, D. P. Siewiorek "Automated Synthesis of Data Paths in Digital Systems" IEEE Trans on CAD, vol. CAD-5, no. 3, pp. 379-395, July 1986.

[6] H. Mecha, M. Fernández, J.F. Tirado, J. Septièn, D. Mozos, K. Olcoz "A Method for Area Estimation of Data-Paths in High Level Synthesis", IEEE Trans on CAD, vol. 15, no. 2, February 1996.

7

EVALUATION OF AN INTEGRATED HIGH-LEVEL SYNTHESIS METHOD

P. Arató, I. Jankovits, Z. Sugár, Sz. Szigeti

Technical University of Budapest,
Department of Process Control,
Hungary

ABSTRACT

Benchmarking in High-Level Synthesis is one of the most critical points nowadays. Since several synthesis and data path allocation methods are published, the comparison between these methods is highly important. The goal of this paper is to summarise the basic steps of the HLS design flow trough the PIPE synthesis tool integrated into the Cadence environment, which has been developed at the Department of Process Control, and to describe some of the most popular HLS benchmarks. This paper also presents the basic definitions of the DFGs elements (functional elements, data connections), which are used as the description of the task. These definitions are indispensable for the correct comparison.

Glossary

Data path - A directed graph representation of data transitions in a problem. Graph nodes are operations, edges are data connections and dependencies.

Execution time - Time needed by an (elementary) operational unit to calculate its output value from its inputs. Denoted by $t(i)$, where i is the number of the operation.

Latency - Time difference between a set of data entering the data path and the output values belonging to that set of input data becoming available on the outputs. (L)

Loop or Recursive loop - A section in a data path executed in an iterative way such that every iteration requires the result of the previous iteration (as initial value) and data from the data path. As the time between successive iterations of the loop may not be smaller than the total of all execution units in the loop (Lr), the loop takes data from the outside at most with the frequency equal to 1/Lr.

Pipelined execution - Feeding a system with a restart time less than total latency is available in some units. Any such unit is executing in an overlapped, pipelined way.

Restart time - The period time of new data input to the system. (R)

Time steps or Time cycles - The time unit in time calculations, often expressed without dimension, i.e. "a time of 3 [time cycles]".

Functional Element (e(i) or Fei):
1.   e(i) is started only after having finished every e(j), for which e(j)⟶e(i) (where e(j) is the predecessor of e(i)) holds.
2.   e(i) requires all its input data during the whole duration time t(i)
3.   e(i) may change its output during the whole duration time t(i)
4.   e(i) holds its actual output stable until its next start.

1   INTRODUCTION

The last few years the High-Level Synthesis has become a very popular research field where several methods were published [1,3,4,6-8,11,13,14]. The common goal of these methods is to find an efficient structure at the RTL level from a behavioral -mostly VHDL or VHDL-like- description with an acceptable computation time. Since the substeps of the HLS are NP-hard problems the question of the processing time is always solved by introducing restrictions which result in near-optimal solutions. The data-flow graph (DFG) synthesis has two basic tasks. First, the scheduling whose goal is to place the functional elements to the control step where the concurency is minimal. Second, the allocation that aims to minimise the number of the needed functional units (adder, multiplier, devider, ALUs or processors). A well-known strategy, and maybe the one that is referred most of the time is the force-directed scheduling [8] which proposes to reach the most efficient structure by balancing the mobilities in the DFG. Another approach to synthesise DFGs are the ILP methods. Generally these methods can handle only single cycle operations [4], but there are some known expansion for multi-cycle operations too [17]. The only thing that all of the mentioned methods accept is that all the problems that can arise during the HLS design flow are solved inside of the DFG. Opposed to this very important point, there are some published CDFG (control data-flow graph) oriented methods where external elements -or signals- are admissible [10,14]. These methods usually lead to a heuristic direction, and this strongly makes the comparison to the optimal solution more difficult (since the optimal solution is unknown).

Another feature of the HLS is the overlapped execution called pipelineing. The goal of the DFG pipelineing is to increase the throughput of the system (reduce the restarting period), while the number of resources is still held on the minimal value. To compare the results of the HLS tools which can handle pipeline restarting period is much more difficult than in the case of non-pipelined ones. Even the basic definitions can be very different in different articles. Some researchers use the notion of 'pipeline stages' [8,10], while others use 'restarting period' [1-3,9-12], in some articles the 'latency' is the period time of new data input to the system, but somewhere else it is called 'restart time'. However, the fundamental problem is in the physical domain of the design. In contrast to the 'old axiom' stating that during the HLS we don't deal with the physical constraints, some of the physical parameters still have to be defined before the design-flow starts. The execution times have to be defined (they are inputs in every methods), and for the comparable results the rate of the speed must be defined too. That is the point where the number of the various views is nearly equal to the number of the articles if these problems are discussed at all. Pipelineing raises some special questions too:

- What to do with recursion? [9]
- How is it possible to handle the conditional branches? [10,11]

The answers have been worked out usually independently without specifying the scheduling or the allocation algorithms, but some approaches [10] use the 'single cycle functional elements' restriction again.

Since so many researchers work in so many ways, in order to solve the questions of HLS it is highly important to find a common base and a common language where a comparison can be made between the different results. In this paper, a technique for DFG synthesis will be presented, also some benchmarks will be shown that might be suitable for testing different HLS tools.

2   PIPE

The PIPE is an educational software tool that has been developed at the Technical University of Budapest, Department of Process Control. The input of the program is an HDL specification of the design and the output is a trade-off between the restarting period and the needed area (the number of the processors) with fixed maximal latency. In the first step the HDL is translated into an inner structure. The translator contains a schemantical checking function. The minimum of the latency is calculated in this step. The ' Variation ' generates all the possible placements for the synchronisation buffers, and pushes it towards to the allocation one by one. The allocation is done for each structure, and the decision about the optimum is just made when all results are known.

■   Scheduling

This algorithm is executed when the difference between the latency minimum and the given latency allows the buffer insertion discussed in [12]. The scheduler schedules the structure for the given restarting period and calculates the synchronisation parameters. While a buffer is cheaper than an other copy from a functional elements, this scheduler algorithm guaranties the cheapest structure which is restartable with the given restarting period.

```
for i (every elements)
if t(i)>=R-1 then c=[t(i)+2/R] ;
    multiply c times ;
    insert a buffer to every input of every copy ;
  else for j (every next elements in data connection)
    if t(i)+t(j)>=R-1 then insert a buffer between e(i) and e(j)
    next j
if e(i) receptor then SYNC [12]
next i
```

■   Variations

The Variations generate all the possible placement of the buffers was inserted to solve the synchronisation problem. This means that all the functional units have mobility in the DFG go through all the time steps between there ASAP

and ALAP position. The best arrangement will result the cheapest final structure after the allocation. If the ith path needed to be delayed by inserting p(i) buffers and the path contains n(i) elements, then the number of the possible variations is: $w(i)=(p(i)+n(i))!/(p(i)!*n(i)!)$
The number of the variations for the whole graph is: $W=\Pi\ w(i)$

```
for i (every path which needed to be delayed)
   for j (every placement between ASAP and ALAP)
   generate a new variation
   next j
next i
```

- Allocation

The allocation algorithm follows the method which was detailed in [1,12]. In this step the program selects the functions which can be slipped into one processor and solves the allocation problem for each group. During the allocation three tasks are executed:

> *Concurency examination*: in this step the allocation checks all functional elements if they are concurrent or not.
> *Generating the maximal compatibility groups.*
> *Coverage*: here the allocation chooses the cheapest set of groups from the maximal compatibility groups.

- Analysis of Conditional Branches

If the DFG contains conditional branches (if-then-else statement), the allocation should be modified. The difference between a condition-detector element and a general element is that just one of the following transfer sequences executes depending on the condition value. Since two parallel branches have EXOR like execution the concurrency examination must be changed. The new formula is [11]:

$$K = [(c(j)*k_h(j)+h)-(c(i)*k_S(i)+s)]*R$$
$$b(i)-b(j)-q(j)\leq K \leq b(i)-b(j)+q(j)$$

where,   b(i), b(j)   : the first start time of e(i) and e(j)
           c(i), c(j)   : the copied number of e(i) and e(j)
           $k_h(i)$, $k_s(i)$: a integer values
           R   : the restarting period
           q(i), q(j)   : the transfer score of e(i) and e(j)
           more detailed definitions in [1].

In case of alternative branches, the K=0 solution doesn't mean concurency for
the functional element pairs.

■   Pipelineing in multi-user sequential recursive loops

Recursive loops are considered to be unavailable to overlapped execution during
the scheduling phase of ASIC design. This is caused by the special nature of
recursive execution: an iterative algorithm may not be fed the next data before
the final result of the previous iteration is ready. In contrast with this fact there
are numerous papers [6,8] which use pipelined recursive loops as benchmarks.
The most popular one is the differential equation solver [8]. The mathematical
definition (1) of this problem shows clearly the recursive nature.

$$x1=x+dx; \ y1=y+dy; \ u1=u-3xudx-3ydx \ (1)$$

In [8] Paulin schedules the DFG (see in 3.4) for overlapped execution without
dealing with the recursion.

There are some notable exceptions, however, to the general case. In a special
type of problems, recursive solutions are needed to calculate values of identical
functions for different processes. From [9], it is known that by using multiple-
process recursive loops, it is possible for separate processes to share the same
resources in such a way that the recursive core in process is realised only once.
It means that the total loop sequence will be divided into smaller parts where
each part can work parallel on a task and the parts rolling through the loop
without breaking the rule of the recursion (pipelined recursive loop). Also from
[9] can be read out the conditions when it is worth using this method:

$$\frac{R}{L} \ll \frac{n-1}{n} \ (2)$$

The right side of (2) shows that the more the restarting period is decreased against the latency, the more efficient structure will be achieved and from the left side it can be seen that as more and more data is introduced to the structure, the result can be more and more efficient.

3 TOOL INTEGRATION INTO COMMERCIAL FRAMEWORKS

State-of-the-art CAD systems should provide more and more aid for the electronic designers. This means that a good design system should support consistent descriptions in all design description domains (system specification, behavioral, structural and physical levels) and integrated tools for the simulations and for the verifications. It is also required to automate the design steps as much as possible to reduce the needed development time and hence the cost of engineering.

Modern CAD systems can accomplish these tasks with providing consistent database management and access, powerful communication protocol for the different kind of tools and user friendly, uniform graphics environment for the user. This complex system is called framework. In most cases it may be extended by programming in a vendor dependent language. Nowadays the simulation procedure is one of the most critical points of the design process. Of course all description domains have to be simulated with the same simulator, if possible. In this case the electronic designer can apply the same test vectors (or with minor modifications) which increases the reliability of simulation and simultaneously reduces the required development time. The framework should provide the back annotation feature to improve the accuracy of the descriptions at the higher abstraction levels. Finally, CAD systems should support the available ASIC technologies (FPGA, VLSI and MCM).

When a new design methodology has been developed, it is recommended to connect to an existing framework environment, unless we wish to create a completely new CAD system. Since in most cases the new tool covers just a few design steps thus probably the extension is the better choice and it requires less effort.

A special type of the extensions is the interfacing. In this case the new CAD tool is not a part of the framework environment, just a standalone application that has a well-defined interface. This interface is usually a standard hardware description language (VHDL or Verilog), a netlist format (EDIF) or physical description file (for VLSI layouts CIF, GSDII, etc.) depending on the abstraction level. Interfacing has several disadvantages. First of all, it does not provide a uniform database management and access. Because of the required data conversion some information may have been lost. Implementation of the simulation method was discussed earlier with the forward and back annotations is also not easy. Because of these reasons, interfacing as an extension method is recommended in the case of small applications that realise a small number of design steps and where these constraints do not cause flexibility loss.

The real integration of a CAD tool does not have these disadvantages but requires much more programming effort from the CAD engineers because of the complexity of the commercial framework products.

High-Level Synthesis Tool Integration to the Design Framework II

One of the most popular frameworks is the Design Framework II. [18] of the Cadence OPUS that is extensible with C or SKILL [19] programming languages. A new high-level synthesis tool has been developed and integrated into this framework. This program provides a user-friendly environment for editing, simulating and synthetising the data-flow graph, as illustrated in Fig. 1. The synthesis procedures cover the algorithms discussed in this paper. The created data-flow graph may be a part of a larger design, its inputs and outputs may be connected to other schematics and HDL descriptions.

Figure 1: Data-flow graph editor

The design flow of high-level synthesis is shown in Fig. 2. The synthesis procedure starts with the data-flow graph that may be entered by using the graphics editor or may be derived from a VHDL description via a VHDL to DFG compiler. This VHDL interface may be useful because the well-known benchmarks are written in VHDL. The graph is stored on the disks in the standard CDBA format (Cadence Database) and a completely new view type is associated with it (*dfg*). The view of the *VHDL* description is derived from the VHDL view type. This description may be simulated by the LeapFrog. simulator. The *nodes* of the data-flow graph are described in three different views. The *node* view contains the symbol and I/O information and is derived from *symbol* view type. The *functional* view describes the behaviour of the operation and it is written in Verilog language [20]. Finally, the *nodeprop* view defines the properties associated with the processor. This information is stored as a set of SKILL statements. The predefined operators are collected in the library *DfgLib*. The user may extend this library by creating the listed views.

After defining the HLS constraints and selecting the type of the scheduling by the user, the synthesis procedures may be started in the given order. If required, the optimiser creates the optimised data-flow graph with the view *dfg.opt*. The scheduler and the allocator can create a cost per performance curve to aid the designer in selecting the run configuration with the most optimal trade-off. Immediately after the scheduling the data- flow graph may be simulated standalone or with the other part of the design by the Verilog-XL simulator. As a final step of synthesis, the module generator creates the controlled data-flow graph (with the view of *cdfg*) and the Verilog RT level description of processors and control logic. The module generator provides additional constraints for RTL synthesiser, based on the original user parameters and the internal structure of the data-flow graph. The next design step is the technology mapping for the available technologies (gate arrays, VLSI, MCM). Using the final verification tools the exact delay times may be extracted from the layout for simulation and back annotation purpose.

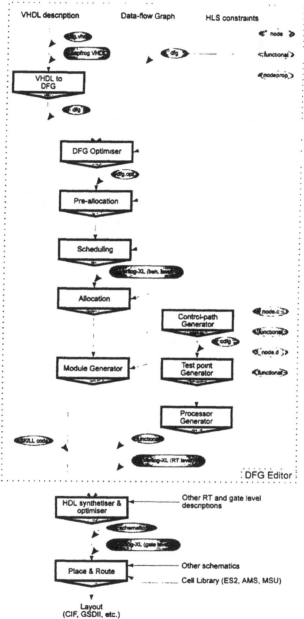

Fig.2. Design flow of HLS tool in the Design Framework II

4 BENCHMARKS

This section contains four very popular benchmarks with their DFGs, VHDL behavioral description and the results that were generated by the PIPE synthesis tool. In the case of Differential Equation Solver, the results were produced in a half automatic way (PIPE cannot handle recursive loops at this momement). The execution times are used in this section (in clock cycles):

$$t(*) = 8 \qquad t(+) = 4 \qquad t(\text{buff.}) = 1$$

4.1 FIR filter

The FIR filter is one of the most popular benchmark that was published in numerous HLS reports [6,7,9,12,15] . On this example it can be seen that how important is the definition of the task. At this moment we don't know a exact method which could transform an 'optimal' DFG from a higher level description (equation, C function ...), since we don't even know what optimal means in the case of a DFG. Comparing the structure a; and c; it is noticeable that the mobilties in the case of structure a; are more than in the case of structure c;, while the latency is less in the structure c; (L(c)=5; L(a)=9). The structure b; is a recursive application. From section 2 (equation 2) it is clear that there is no worth pipelining in this case.

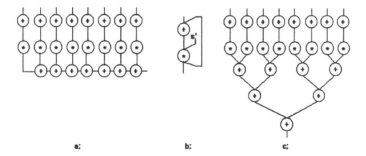

a;   b;   c;

The VHDL behavioral description of structure c is:

```
PACKAGE Global IS
        TYPE Vektor IS ARRAY (NATURAL RANGE <>) OF INTEGER;
        FUNCTION "+" (L,R : Vektor) RETURN Vektor;
END Global;

PACKAGE BODY Global IS

FUNCTION "+" (L,R : Vektor) RETURN Vektor is
VARIABLE result : Vektor(L'Range);
begin
        for i in L'Range loop
                result(i) := L(i) + R(i);
        end loop;
        return result;
end;

END Global;

USE work.Global.ALL;

ENTITY Fir_szuro IS
PORT ( x: IN Vektor; y: OUT INTEGER );
END Fir_szuro;

ARCHITECTURE vis OF Fir\_szuro IS
        SIGNAL  a,b : Vektor(1 TO 8);
        SIGNAL  c   : Vektor(1 TO 4);
        SIGNAL  d   : Vektor(1 TO 2);
        CONSTANT w: Vektor(1 TO 8) := (others$=>$ 1);
BEGIN
PROCESS (x)
        VARIABLE i,j,k : INTEGER;
BEGIN
        FOR i IN 1 TO 8 LOOP
                a(i)<= x(i-1) + x(i) AFTER 4 ns;
                b(i) <= a(i) * w(i) AFTER 8 ns;
                IF (i MOD 2) = 0 THEN
                        j := i/2;
                        c(j)<= b(i-1) + b(i) AFTER 4 ns;
```

```
                   END IF;
                   IF (i MOD 4) = 0 THEN
                           k := i/4;
                           d(k)<= c(i/2-1) + c(i/2) AFTER 4 ns;
                   END IF;
           END LOOP
END PROCESS;
           y <= d(1) + d(2) AFTER 4 ns;
END vis;
```

The PIPE results:

R:	7	9	11	13	15	30
a;						
*	16	8	8	8	6	8
+	14	14	14	12	10	8
buf	83	52	52	40	29	0
c;						
*	16	16	8	8	8	8
+	15	15	14	14	14	8
buf	44	22	16	0	0	0

4.2   Expansion of 3*3 Determinant

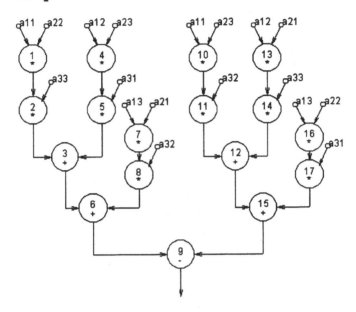

The VHDL behavioral description:

```
USE work.Global.ALL;

ENTITY graph IS
    GENERIC (mt:time:=8 ns; at:time:=4 ns; st:time:=4 ns);
    PORT(a11,a12,a13,a21,a22,a23,a31,a32,a33:in Real;
    kim:out Real);
END graph;

ARCHITECTURE Viselk OF graph IS
    SIGNAL m1, m2 ,m3, m4, m5, m6, m7, m8, m9, m10: Real;
    SIGNAL m11, m12, a1, a2, a3, a4: Real;
BEGIN
    m1 <= a11*a22 After mt;
    m2 <= a12*a23 After mt;
    m3 <= a13*a21 After mt;
    m4 <= a11*a23 After mt;
```

```
m5  <= a12*a21 After mt;
m6  <= a13*a22 After mt;
m7  <= m1*a33  After mt;
m8  <= m2*a31  After mt;
m9  <= m3*a32  After mt;
m10<= m4*a32  After mt;
m11<= m5*a33  After mt;
m12<= m6*a31  After mt;

a1  <= m7+m8   After at;
a2  <= m10+m11 After at;
a3  <= a1+m9   After at;
a4  <= a2+m12  After at;

kim <= a3-a4 After st;
END Viselk;
```

R:	14	16	18	20	22
AL	17	17	17	17	17
buf	68	68	56	56	56

4.3   Differential Equation Solver

The description of the DES is a small fixed-point calculation loop. The algorithm tries to numerically solve the equation:

$$y'' + 3xy' + 3y = 0$$

Here, u is assumed to represent dy/dx or y'. dx is approximated as x1-x. Similarly, dy = y1-y and du = u1-u. The value 'a' provides the number of times the numerical loop is executed. u1, x1 and y1 represent the new values of u, x and y. Thus, x1 = x+dx, y1 = udx+y, u1 = u-3xudx-3ydx. The behavior executes by loading the initial values of x, y, u, dx, and a .

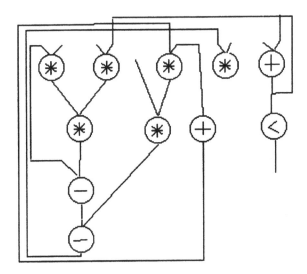

The VHDL behavioral description of the DES:

```
USE work.Global.all;

entity diffeq is
   port (Xinport: in integer;
         Xoutport: out integer;
         DXport: in integer;
         Aport: in integer;
         Yinport: in integer;
         Youtport: out integer;
         Uinport: in integer;
         Uoutport: out integer);
end diffeq;

architecture diffeq of diffeq is

begin

P1 : process (Aport, DXport, Xinport, Yinport, Uinport)
```

```
   variable x_var,y_var,u_var, a_var, dx_var: integer ;
   variable x1, y1, t1,t2,t3,t4,t5,t6: integer ;

begin

x_var:=Xinport;a_var:=Aport;dx_var:=DXport;y_var:=Yinport;
u_var:=Uinport;

   while (x_var < a_var) loop

       t1 := u_var * dx_var;
       t2 := 3 * x_var;
       t3 := 3 * y_var;
       t4 := t1 * t2;
       t5 := dx_var * t3;
       t6 := u_var - t4;

       u_var := t6 - t5;
       y1 := u_var * dx_var;
       y_var := y_var + y1;
       x_var := x_var + dx_var;


   end loop;

  Xoutport <= x_var;
  Youtport <= y_var;
  Uoutport <= u_var;

end process P1;

end diffeq;
```

The PIPE results (the graph was tuned with opened loop, and the number of the buffers in the feedback-path was calculated manualy):

R:	7	9	11	13	15	30
AL	14	13	9	7	8	6
buf	125	92	51	41	35	3

5 SUMMARY

A HLS method and the steps of a evaluation method were described. In this paper four popular benchmarks was shown that could be suitable to compare different HLS tools. The DES can be usefull for the application can handle recursions. The published solutions were generated by the PIPE design tool. The PIPE can schedule and allocate structures which do not contain conditional branches (to test this part benchmarks can be found in [10]). Another important property of the PIPE is that it can allocate just the same type of processors (the functions in the behavioral level and the processors in the RTL level are the same). It is also very important to give the exact definition of the functional element (timing constraints, if it contain storage element or not, ... see in Glossary) to have correct comparison.

6 ACKNOWLEDGEMENTS

This work was supported in part by the European Office of Aerospace Research and Development, Contract No. 170895W0281, in part by COPERNICUS, Contract No. CP-940453 and in part by OTKA , Contract No. T 017236.

7 REFERENCES

[1] Péter Arató, Andrzej Rucinski, Robert Davis, Roy Torbert, István Bires, A High- Level Datapath Synthesis Method for Pipelined Structures, Microelectronics Journal 25,1994.

[2] Péter Arató, A Data-flow Model and Method for Optimizing the Pipeline Restarting Period, Proc. of The Eighth Symposium on Microcomputer and Microprocessor Applications, 1994.

[3] Péter Arató, Andrzej Rucinski, Istvan Jankovits,Time Scaled High-Level Synthesis for Pipelined Data-flow Structures, Proceedings of ATW'94.

[4] Cheng-Tsung Hwang, Jiahn-Hurng Lee, Yu-Chin Hsu, A Formal Approach to the Scheduling Problem in High Level Synthesis, IEEE Transactions on Computer- Aided Design, Vol. 10. No. 4, April 1991.

[5] High-Level VLSI Synthesis, Edited by Raul Camposano & Wayne Wolf, Kulwer Academic Publisher, 1991.

[6] Wolfgang Rosenstiel, Heinrich Kramer, Scheduling and Assignment in High Level Synthesis*

[7] PISYN- High-Level Synthesis of Application Specific Pipelined Hardware, Albert E. Casavant, Ki Soo Hwang, Kristen N. McNall*

[8] Pierre G. Paulin, John P. Knight, Force-directed Scheduling for the Behavioral Synthesis of ASIC's, IEEE Transactions on Computer-Aided Design, 1989/6.

[9] Istvan Jankovits, Tamás Visegrádi, Pipelined execution in multi-user sequential recursive loops, Periodica Politechnika, Accepted.

[10] Ivan P. Radivojevic, Forrest Brewer: Analysis of Conditional Resource Sharing using a Guard-based Control Representation, Computer Hardware and Design, October 1994.

[11] István Béres, Ph.D. theses, Technical University Budapest.

[12] István Jankovits, A Scheduling and Allocation Method Based on a Time-Scaled Algorithm, Proc. of The Eighth Symposium on Microcomputer and Microprocessor Applications, 1994.

[13] Gábor Paller, Rafael, an Intelligent, Multi-Target Signal-Flow Compiler, Ph.D. theses, Technical University Budapest.

[14] Ivan P. Radivojevic, Forrest Brewer, Symbolic Scheduling Techniques, Computer Hardware and Design, 1994.

[15] Alice C. Parker, Kayhan K—c—kcakar, Shiv Prakash, Jen-Pin Weng, Unified System Construction*

[16] R. Camposano, R. A. Bergamaschi, C. E. Haynes, M. Payer, S. M. Wu: The IBM High-Level Synthesis System*

[17] Yu-Chin Hsu, Youn-Long Lin, High-Level Synthesis in the Theda System*

[18] Design Framework II Reference Manual 4.3, March 1994.

[19] SKILL Language Reference Manual 4.3, March 1994.

[20] Verilog-XL Reference Manual 2.0, March 1994.

*˙ High-Level VLSI Synthesis, Edited by Raul Camposano & Wayne Wolf, Kluwer Academic Publisher, 1991

8

COMBINATORIAL CRITERIA OVER GRAPHS OF SPECIFICATION TO DECIDE SYNTHESIS BY SEQUENTIAL CIRCUITS

Y. P. Tison, P. Simonnet

Centre de Mathématiques et de Calcul Scientifique,
Université de Corse, B.P. 52, 20250 Corte, France

ABSTRACT

Here we present some algorithms which decide, for a given functional specification, whether the function is continuous and whether the function is sequential. When the specification is synchronous (i.e the graph of the function is realized by a synchronous automata) then these two notions coincide with asynchronous sequential functions with bounded delay. We give an example where Büchi's synthesis by a synchronous sequential function is not possible, but synthesis by an asynchronous sequential function with bounded delay is possible. When the specification is asynchronous, we present an example of a continuous but not sequential function, and we give a sufficient criterion to prove that a function is not sequential.

1 INTRODUCTION

Links between finite automata and sequential circuits are well known. Two aspects are interesting in sequential circuits: the verification of a specification and the synthesis of a specification. The importance given to them was one of the reasons which motivated people to study automata over ω-words [5].

In 1962 Büchi proved the decidability of S1S, the monadic theory of one successor [3]. Büchi's proof shows that a S1S specification can be seen as an automaton over ω-words. Therefore the verification of a S1S specification of a sequential circuit reduces to testing the inclusion between languages recognized by automata over ω-words, which is a decidable property.

109

In the 80's, for non-terminating reactive finite state programs this result was extended to several temporal logic system in place of S1S. Now these verification procedures are implemented. For example, the synchronous programming language Esterel has a module Tempest which verifies the temporal logic property of programs. See [9] for an overview of program verification (model checking).

In 1969 Büchi and Landweber, in the context of finite state games, showed that the synthesis problem for a S1S specification is decidable [4]. In the positive case, the algorithm provides a deterministic finite automaton which outputs letters and realizes the specification given. There are cases where the synthesis is not possible with such an automaton, but is possible by a deterministic finite automaton which outputs words instead of letters. This is our starting point of investigation. Our specifications will be graphs of functions, whether defined in S1S (synchronous specification), or defined by non-deterministic automata which output words (asynchronous specification).

In the synchronous case the synthesis by a finite state device will be possible if and only if the function is continuous. In this case, a deterministic automaton which outputs words with bounded delay between the input and the output realizes the specification. We present a combinatorial test over the graph of the function which decides the continuity of the function.

In the asynchronous case, we give examples of continuous functions which can not be synthetized by a finite state device. We present a combinatorial test over the graph of the specification which implies the non-realizability by a finite state device. Our criterion is an extension to infinite word transducers of Choffrut's criterion. We do not yet know if this criterion is necessary. The continuous functions given by asynchronous specifications are recursive functions. Our example of a continuous function which can not be synthetized by a finite state device can be implemented with a stack.

In the second section, we state some well-known definitions that we will need: Büchi automata, Muller automata, S1S theory, transducers and Büchi Landweber theorem. We need also elements of topology because, in our case, a function is continuous if and only if its graph is closed. Moreover, the natural classification of automata over ω-words is done using topological methods. In the third section, we study the synchronous case and we present our criterion for continuous functions. In the fourth section we treat the asynchronous case and we give our sufficient criterion.

2 DEFINITIONS

We present here some definitions we will use later.

2.1 Infinite words

Concerning the sets: $\mathcal{P}(Q)$ denotes the power set of a set Q. The set of integers is denoted by $\mathbb{N}$.

Let X be a finite alphabet. An infinite word over X, called a ω-word, can be seen as a mapping $\alpha : \mathbb{N} \to \mathbb{X}$. So a ω-word α is a sequence $\alpha = \alpha(0)\alpha(1)\alpha(2) \ldots \alpha(n) \ldots$ where $\alpha(n)$ denotes the nth letter of α and $\alpha[n]$ denotes the finite left factor of length n of α. We will define $|u|$ to mean the length of the finite word u, so $|\alpha[n]| = n$.

Over X, the set of finite words is denoted by X^* and the set of infinite words is denoted by X^ω. The empty word is noted ε.

We will often employ examples concerning characteristic functions of subsets of $\mathbb{N}$. A characteristic function of a set can be seen as an ω-word, with the least terms at the beginning of the word. For example $\alpha = 10111001000\ldots$ represents the set $\{0, 2, 3, 4, 7\}$, $\alpha = 0^\omega$ is the empty set $\emptyset$ and $\alpha = 1^\omega$ is $\mathbb{N}$ in its entirety.

2.2 Topology

Let be X a finite alphabet. In X^ω we will consider the natural product topology. This topology is also defined by the distance d

$d : X^\omega \times X^\omega \to \mathbb{R}^+$ and

$$d(\alpha, \beta) = \begin{cases} 0 & \text{if } \alpha = \beta; \\ \frac{1}{2^n} & \text{with } n = \min\{k/\ \alpha(k) \neq \beta(k)\}. \end{cases}$$

With this topology X^ω is a compact metric space.

We define here the first level of the Borel hierarchy in X^ω. We use the logicians notation as in Moschovakis's book. The exponent equal to zero corresponds to the first order quantifications. The suffix gives the number of quantifier alternation. "Σ" denotes the "$\exists$" quantification. "Π" denotes the "$\forall$" quantification.

$\mathbf{\Pi}_1^0$ is the class of closed sets;

$\mathbf{\Sigma}_1^0$ is the class of open sets;

$\mathbf{\Pi}_2^0$ is the class of denumerable intersections of open sets;

$\mathbf{\Sigma}_2^0$ is the class of denumerable unions of closed sets.

Theorem 2.1. *Let be X^ω and Y^ω two metrisable compacts, and let be G the graph of the function $f : X^\omega \to Y^\omega$, then*

f *is continuous* $\iff$ G *is closed.*

Definition 2.2. A function $f : X^\omega \to Y^\omega$ is continuous in x_0 if

$$\forall n \geq 0, \exists m \geq 0 / \ \forall x \Big((x_0[m] = x[m]) \implies (f(x_0)[n] = f(x)[n]) \Big).$$

Theorem 2.3. *The set of the continuous points of a function is a $\mathbf{\Pi}_2^0$ set.*

2.3   Büchi automata

Here we review the Büchi automata which correspond to usual finite automata applied to infinite words.

Definition 2.4. A Büchi automaton $\mathcal{A} = \langle X, Q, I, \delta, F \rangle$ consists of a finite alphabet X, a finite set Q of states, a set of initial states $I \subseteq Q$, a set of final states $F \subseteq Q$ and a next state function

$$\delta : Q \times X \to \mathcal{P}(Q) \text{ and } \delta(p, x) = \{q \in Q / \ q \in \delta(p, x)\}.$$

A Büchi automaton $\mathcal{A}$ is deterministic if the set of initial states is reduced to only one element q_0 and if for each letter and each state there is only one transition

$$\begin{cases} I = \{q_0\}; \\ \forall x \in X, \forall p \in Q, \ \text{Card}\Big(q \in Q / \ q = \delta(p, x)\Big) = 1. \end{cases}$$

The run of a Büchi automaton $\mathcal{A}$ with an ω-word $\alpha = \alpha(0)\alpha(1)\alpha(2) \ldots \alpha(n) \ldots$ gives an infinite sequence of states $c(\alpha) = q_0 q_1 q_2 \ldots q_n \ldots$ with

$$\begin{cases} q_1 \in I; \\ q_{i+1} \in \delta(q_i, \alpha(i)). \end{cases}$$

We define the set of states which appear infinitely often in $c(\alpha)$ by

$$\text{Inf}(c(\alpha)) = \{p \in Q / \ \text{Card}(i / \ q_i = p) = +\infty\}.$$

Definition 2.5. Let a Büchi automaton $\mathcal{A} = \langle X, Q, I, \delta, F \rangle$. An ω-word $\alpha \in X^\omega$ is accepted by $\mathcal{A}$ if there is a run $c(\alpha)$ such that $\text{Inf}(c(\alpha))$ contains at least one final state. The ω-language recognized by $\mathcal{A}$, that is the set of ω-words accepted by $\mathcal{A}$, is

$$L_\omega(\mathcal{A}) = \{\alpha \in X^\omega \,/\, \exists c(\alpha), \mathrm{Inf}(c(\alpha)) \cap F \neq \emptyset\}.$$

Example 2.1. Let the Büchi automaton $\mathcal{A} = \langle X, Q, I, \delta, F\rangle$ (fig.1) where $X = \{0, 1\}, Q = \{q_0, q_1\}, I = \{q_0\}, F = \{q_0\}$. This automaton recognizes the set containing the word $\alpha = 0^\omega$. This word is the characteristic function of the empty set over $\mathbb{N}$. It is a $\mathbf{\Pi_1^0}$ set.

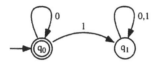

Figure 1

Example 2.2. Let the Büchi automaton $\mathcal{A} = \langle X, Q, I, \delta, F\rangle$ (fig.2) where $X = \{0, 1\}, Q = \{q_0, q_1\}, I = \{q_0\}, F = \{q_1\}$. This automaton recognizes the set of words which have at least one 1. This set is also the set of characteristic functions of non-empty subsets of $\mathbb{N}$. It is a $\mathbf{\Sigma_1^0}$ set.

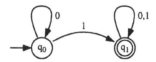

Figure 2

Example 2.3. Let the Büchi automaton $\mathcal{A} = \langle X, Q, I, \delta, F\rangle$ (fig.3) where $X = \{0, 1\}, Q = \{q_0, q_1\}, I = \{q_0\}, F = \{q_1\}$. This automaton recognizes the set of words which have a infinite number of 1s. This set is the set of characteristic functions of infinite subsets of $\mathbb{N}$. It is a $\mathbf{\Pi_2^0}$ set.

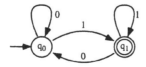

Figure 3

Example 2.4. Let the Büchi automaton $\mathcal{A} = \langle X, Q, I, \delta, F \rangle$ (fig.4) where $X = \{0,1\}, Q = \{q_0, q_1, q_2, q_3, q_4\}, I = \{q_0, q_2\}, F = \{q_0, q_3\}$. This automaton recognizes the set of words which have a finite number of 1s. This set is the set of characteristic functions of finite subsets of $\mathbb{N}$. It is a Σ_2^0 set.

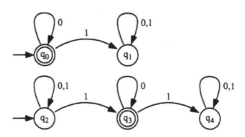

Figure 4

Definition 2.6. A Büchi automaton $\mathcal{A} = \langle X, Q, I, \delta, F \rangle$ is unambiguous if for each ω-word α of $L_\omega(\mathcal{A})$, there is only one sequence $c(\alpha)$ which makes α accepted

$$\forall \alpha \in L_\omega(\mathcal{A}), \ \mathrm{Card}\Big(c(\alpha)/\ \mathrm{Inf}(c(\alpha)) \cap F \neq \emptyset\Big) = 1.$$

Example 2.5. Let the Büchi automaton $\mathcal{A} = \langle X, Q, I, \delta, F \rangle$ (fig.5) where $X = \{0,1\}, Q = \{q_0, q_1, q_2, q_3\}, I = \{q_0\}, F = \{q_1\}$. This automaton recognizes the words beginning by 0 with at least one 1. It is ambiguous because the words whose first two letters are 01 give two different runs.

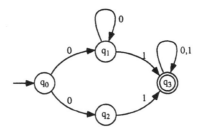

Figure 5

Example 2.6. Let the Büchi automaton $\mathcal{A} = \langle X, Q, I, \delta, F \rangle$ (fig.6) where $X = \{0,1\}, Q = \{q_0, q_1, q_2\}, I = \{q_0\}, F = \{q_1\}$. This automaton recognizes the words containing a finite non-zero number of 1s. This is a non deterministic but unambiguous Büchi automaton.

Figure 6

2.4 Muller automata

We introduce here the Muller automata.

Definition 2.7. A deterministic Muller automaton $\mathcal{A} = \langle X, Q, q_0, \delta, \mathcal{F} \rangle$ consists of a finite alphabet X, a finite set Q of states, an initial state $q_0 \in Q$, a set of subsets of final states $\mathcal{F} \subseteq \mathcal{P}(Q)$ and a next state function

$\delta : Q \times X \to Q$ and

$\delta(p, x) = q.$

By induction we extend this transition function over words of X^* setting

$\delta : Q \times X^* \to Q$ and

$\begin{cases} \delta(p, \varepsilon) = p; \\ \delta(p, ux) = \delta(\delta(p, u), x) \text{ for } u \in X^*, x \in X. \end{cases}$

A deterministic Muller automaton has only one transition for each letter and for each state

$\forall x \in X, \forall p \in Q, \ \mathrm{Card}\left(q/\ \delta(p, x) = q\right) = 1.$

Definition 2.8. Let a deterministic Muller automaton $\mathcal{A} = \langle X, Q, q_0, \delta, \mathcal{F} \rangle$. An ω-word $\alpha \in X^\omega$ is accepted by $\mathcal{A}$ if there is a run $c(\alpha)$ such that the set of states appearing infinitely often is exactly one of the sets describe in $\mathcal{F}$. The ω-language recognized by $\mathcal{A}$, that is the set of ω-words accepted by $\mathcal{A}$, is

$L_\omega(\mathcal{A}) = \{\alpha \in X^\omega /\ \exists c(\alpha), \mathrm{Inf}(c(\alpha)) \in \mathcal{F}\}.$

Example 2.7. Let a deterministic Muller automaton $\mathcal{A} = \langle X, Q, q_0, \delta, \mathcal{F} \rangle$ (fig.7) where $X = \{0, 1\}, Q = \{q_0, q_1\}, \mathcal{F} = \{\{q_0, q_1\}, \{q_1\}\}$. This Muller automaton recognizes the words with a infinite number of 1s.

Theorem 2.9. *Let be $A \subseteq X^\omega$. The following conditions are equivalent:*

i) A is recognized by a non deterministic Büchi automaton;

ii) A is recognized by a deterministic Muller automaton.

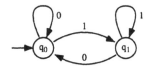

Figure 7

Remark 2.10. It is easy to see that the family of sets recognized by deterministic Muller automata is a boolean algebra. Also, the family of sets recognized by non deterministic Büchi automata is closed under projection.

Definition 2.11. We call *Auto* the class of sets recognized by automata.

Lemma 2.12. *If the language recognized by an automaton $\mathcal{A}$ is not empty then this language contains at least one infinitely periodic word*

$$\left(L(\mathcal{A}) \neq \emptyset\right) \Rightarrow \left(\exists \alpha \in L(\mathcal{A}), \exists u, v \in X^*, \exists q \in Q/ \right.$$
$$\left. \alpha = uv^\omega, q \in \delta(q_0, u), q \in \delta(q, v)\right).$$

Lemma 2.13. *The emptiness problem for automata is decidable.*

To say whether a language $L(\mathcal{A})$ is empty or not is decidable. It suffices to test if $L(\mathcal{A})$ contains at least one infinitely periodic-type ω-word. The construction of one successful word α will be given in finite time.

Corollary 2.14. *The inclusion problem between sets recognized by Büchi automaton is decidable.*

2.5   Landweber characterization

Let $\mathcal{A} = \langle X, Q, I, \delta, F \rangle$ be a deterministic automaton. One can define other acceptance conditions. Let $c \in Q^\omega$ be a run of $\mathcal{A}$

  c is an accepting run if $\exists i \; c(i) \in F$ ($\Sigma_1^0 - condition$);

  c is an accepting run if $\forall i \; c(i) \in F$ ($\Pi_1^0 - condition$);

  c is an accepting run if $\exists j \; \forall i \geq j \; c(i) \in F$ ($\Sigma_2^0 - condition$);

  c is an accepting run if $\forall j \; \exists i \geq j \; c(i) \in F$ ($\Pi_2^0 - condition$).

$\Pi_1^0(Auto)$ is the class of sets recognized by deterministic automata with a $\Pi_1^0 - condition$.

$\Sigma_1^0(Auto)$ is the class of sets recognized by deterministic automata with a $\Sigma_1^0 - condition$.
$\Pi_2^0(Auto)$ is the class of sets recognized by deterministic automata with a $\Pi_2^0 - condition$.
$\Sigma_2^0(Auto)$ is the class of sets recognized by deterministic automata with a $\Sigma_2^0 - condition$.

Remark 2.15. We have the following inclusions:

$$\Pi_1^0(Auto) \subset \mathbf{\Pi}_1^0,$$
$$\Sigma_1^0(Auto) \subset \mathbf{\Sigma}_1^0,$$
$$\Pi_2^0(Auto) \subset \mathbf{\Pi}_2^0,$$
$$\Sigma_2^0(Auto) \subset \mathbf{\Sigma}_2^0.$$

Remark 2.16. $\Pi_2^0(Auto)$ is the class of sets recognized by deterministic Büchi automata.

Theorem 2.17. *(Landweber)*

  i) $\mathbf{\Pi}_1^0 \cap Auto = \Pi_1^0(Auto);$
  ii) $\mathbf{\Sigma}_1^0 \cap Auto = \Sigma_1^0(Auto);$
  iii) $\mathbf{\Pi}_2^0 \cap Auto = \Pi_2^0(Auto);$
  iv) $\mathbf{\Sigma}_2^0 \cap Auto = \Sigma_2^0(Auto).$

This theorem says for example that if a set L is recognized by a non deterministic Büchi automaton and L is a $\mathbf{\Pi}_2^0$ set then we can construct a deterministic Büchi automaton which recognizes L.

Remark 2.18. Landweber gives an algorithm to decide which is the class of a set recognized by a Muller automaton.

2.6 S1S theory

Büchi used sequential automata to prove decidability of the monadic second-order theory of natural numbers with the successor relation which is called, for short, the second-order theory of one successor, or S1S. The variables of S1S range over sets of natural numbers. S1S atomic formulas have a form $A \subseteq B$ or $Succ(A, B)$. The latter means that there is a natural number n with $A = \{n\}$, $B = \{n + 1\}$. Other S1S formulas are built from S1S atomic formulas using conjunction, disjunction, negation and the existential quantifier. Every set A of natural numbers can be identified with its characteristic function, i.e. $A(n) = 1$ if $n \in A$, and $A(n) = 0$ otherwise. For any natural number m, let Σ_m be the direct product of m copies of the set $\{0, 1\}$.

Theorem 2.19. *For every S1S formula ϕ with m variables there is a Büchi Σ_m-automaton $\mathcal{A}$ such that for all sets $A_1, \ldots, A_m$ of natural numbers, $\phi(A_1, \ldots, A_m)$ holds iff $\mathcal{A}$ accepts the Σ_m-sequence*

$$A_1(0), \ldots, A_m(0), A_1(1), \ldots, A_m(1), A_1(2), \ldots, A_m(2). \ [3]$$

The desired automaton $\mathcal{A}$ is constructed by induction on ϕ. The atomic case and the cases of conjunction, disjunction and the existentials quantifiers are easy. To handle the negation it suffices to use the equivalence between non-deterministic Büchi automata and deterministic Muller automata (remark 2.10).

It is easy to see that sets recognized by automata are defined by S1S formulas.

The one-element set $\{x\}$ is given by the automaton $\mathcal{A} = \langle X, Q, q_0, \delta, F \rangle$ (fig.8) where $X = \{0, 1\}, Q = \{q_0, q_1, q_2\}, F = \{q_1\}$.

Figure 8

The successor $\mathrm{Succ}(x, y)$ meaning "*x is the successor of y*" is given by the automaton with two input tapes $\mathcal{A} = \langle X, Q, q_0, \delta, F \rangle$ (fig.9) where $X = \{0, 1\}^2$, $Q = \{q_0, q_1, q_2, q_3\}, F = \{q_2\}$.

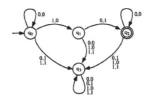

Figure 9

The inclusion $A \subset B$ is given by the automaton with two input tapes $\mathcal{A} = \langle X, Q, q_0, \delta, F \rangle$ (fig.10) where $X = \{0, 1\}^2, Q = \{q_0, q_1\}, F = \{q_0\}$.

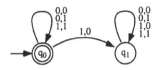

Figure 10

2.7 Transducers and functions realized by automata

Definition 2.20. A morphism φ is a function from X^* into Y^* such that

$$\varphi(\varepsilon) = \varepsilon \text{ and } \varphi(uv) = \varphi(u)\varphi(v).$$

It suffices to define φ on letters of X. If $\forall a \in X$ $\varphi(a) \neq \varepsilon$ then we can extend φ into a function $\varphi : X^\omega \to Y^\omega$.

Definition 2.21. A sequential transducer $\mathcal{T} = \langle X, Y, Q, q_0, \delta, \sigma \rangle$ is a deterministic automaton with an output function. It consists of a finite input alphabet X, a finite output alphabet Y, a finite set of states Q, an initial state q_0, a next state function δ and an output function σ

$$\delta : Q \times X \to Q, \ \sigma : Q \times X \to Y^*.$$

We extend the next state and the output functions over words of X^* by

$\delta : Q \times X^* \to Q$ with

$\delta(q, \varepsilon) = \varepsilon$ and $\delta(q, ux) = \delta(\delta(q, u), x), u \in X^*, x \in X;$

$\sigma : Q \times X^* \to Y^*$ with

$\sigma(q, \varepsilon) = q$ and $\sigma(q, ux) = \sigma(q, u)\sigma(\delta(q, u), x).$

A sequential transducer defines a function $f : X^* \to Y^*$ with $f(u) = \sigma(q_0, u)$.

We call a transition $p \xrightarrow{u} q$ a sequence of states $pp_1p_2 \ldots p_n q$ given by the run of a word $u \in X^*$ from the state p. A loop will be a transition with $q = p$ so $p \xrightarrow{u} p$.

If there is no loop $p \xrightarrow{u} p$ with an output word equal to ε, we can extend f over ω-words $f : X^\omega \to Y^\omega$ with

$$f(\alpha) = \sigma(q_0, \alpha) = \lim_{n \to \infty} (\sigma(q_0, \alpha[n])).$$

If there is a loop $p \xrightarrow{u} p$ with an output word equal to ε, f will be a partial function $f : X^\omega \to Y^\omega$ whose domain is recognized by deterministic Muller automata.

Definition 2.22. A function $f : X^\omega \to Y^\omega$ is sequential if it is realized by a sequential transducer.

Remark 2.23. Any morphism is a sequential function.

Example 2.8. Let f be the multiplication by 2 of an integer $f : \mathbb{N} \to \mathbb{N}$ and $f(x) = 2x$. An integer is defined as a singleton over $\mathbb{N}$. Then its characteristic function is an ω-word containing only one 1. Let $g : \{0,1\}^{\mathbb{N}} \to \{0,1\}^{\mathbb{N}}$ be a function which doubles every 0 before the first 1. If we reduce its domain to the words containing at most one 1 we obtain the multiplication function f in binary code and $g(0^n 10^\omega) = 0^{2n} 10^\omega$. This function g is realized by the sequential transducer (fig.11).

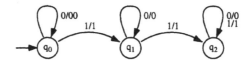

Figure 11

Definition 2.24. A 1-sequential transducer $\mathcal{T}$ is a sequential transducer with the output function reduced to

$$\sigma : Q \times X \to Y.$$

Remark 2.25. The function g presented in (fig.11) can not be realized by a 1-sequential transducer. This can be proved using the pumping lemma. An another way to see it is that its graph can not be defined in S1S because if we add the function $f(x) = 2x$ to the language of S1S the theory becomes undecidable [23].

Definition 2.26. A sequential transducer has a bounded delay iff the length difference between the input and the output labels of every loop is 0.

Example 2.9. The sequential transducer (fig.12) has a bounded delay.

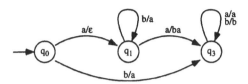

Figure 12

Definition 2.27. A Büchi transducer $\mathcal{T} = \langle X, Y, Q, I, E, F \rangle$ consists of a finite input alphabet X, a finite output alphabet Y, a finite set of states Q, a set of initial states $I \subseteq Q$, a set of final states $F \subseteq Q$, a finite set of transitions $E \subseteq Q \times X \times Y^* \times Q$.

Remark 2.28. If we forget the output we have a non-deterministic Büchi automaton.

Definition 2.29. To normalize a Büchi transducer consists in reducing the set of initial states to an initial state q_0, retaining from the set of states Q which are accessible from q_0, retaining the states from which an ω-word can be recognized.

From now on we will consider only the normalized transducers.

Definition 2.30. An unambiguous Büchi transducer $\mathcal{T}$ is a Büchi transducer such that for each ω-word of $Dom(\mathcal{T})$ there is only one acceptant calculus.

Remark 2.31. With a normalized unambiguous Büchi transducer, $\forall p, q \in Q$, $\forall u \in X^*$ there exists at most one transition $p \xrightarrow{u} q$.

So with the output of any transition we can define the output function σ : $Q \times X \times Q \to Y^*$. If the transition $p \xrightarrow{u} q$ exists then $\sigma(p, u, q)$ is equal to the concatenation of each output present in the transition. If a transition $p \xrightarrow{u} q$ does not exist we have $\sigma(p, u, q) = 0$.

If there exists no loop $p \xrightarrow{u} p$ such that $\sigma(p, u, p) = \varepsilon$ then the transducer defines a partial function $f : Dom(\mathcal{T}) \to Y^\omega$. If $\alpha \in Dom(\mathcal{T})$ there is a unique accepting run $c(\alpha) = q_0 q_1 q_2 \ldots$. The output of α, $f(\alpha) = \sigma(q_0, \alpha(0), q_1)$ $\sigma(q_1, \alpha(1), q_2) \ldots$, is the concatenation of the outputs of the transitions. If there exists a loop such that $\sigma(p, u, p) = \varepsilon$ some infinite words are transformed into finite words. To eliminate these words we reduce the input domain to another one still recognized by Büchi automata.

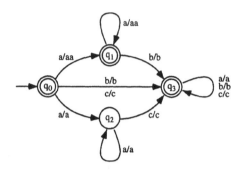

Figure 13

Example 2.10. The Büchi transducer of (fig.13) is unambiguous.

Definition 2.32. A 1-unambiguous Büchi transducer $\mathcal{T}$ is an unambiguous Büchi with the output function reduced to

$$\sigma : Q \times X \times Q \to Y.$$

2.8 Büchi Landweber theorem

Definition 2.33. An infinite game with two players I and II consists of two finite alphabets X and Y, a subset $S \subset (X \times Y)^\omega$. I plays letters from X, II plays letters from Y. Player I constructs an infinite word α. Player II constructs an infinite word β. Player II wins if $(\alpha, \beta) \in S$.

A strategy for player I is a function $f : Y^* \to X$. A strategy for player II is a function $g : X^+ \to Y$ $(X^+ = X^* - \{\varepsilon\})$.

A strategy is winning for I if $\forall \beta (f(\beta), \beta) \notin S$. A strategy is winning for II if $\forall \alpha (\alpha, g(\alpha)) \in S$.

Theorem 2.34. *If $S \subset (X \times Y)^\mathbb{N}$ is recognized by a Muller automaton then one of the players has a winning strategy given by a 1-sequential function (for I it is a Moore automaton). There is an algorithm which given S decides which player has a winning strategy and constructs for this player a 1-sequential function which is a winning strategy.*

See [23] for a proof. In particular if II has a winning strategy then he has a 1-sequential winning strategy. The theorem solves the synthesis problem for

S1S specification. It is equivalent to the decidability of the emptiness problem for infinite tree automata, which is used to prove the decidability of S2S, the monadic theory of the binary tree (two successors). S2S is a powerful tool to prove decidability results and is usually used in the theory of concurrence to express properties of programs. The satisfiability problem for several modal logic like computation tree logic or branching time logic can be translated into a S2S formula (see [2]). Landweber's characterizations and decidability results can be proved easily by the previous theorem.

3   SYNCHRONOUS FUNCTIONS

Proposition 3.1. *(Arnold) For each Muller automata we can construct an unambiguous Büchi automata which recognizes the same language.*

Corollary 3.2. *Let be a function $f : X^\omega \to Y^\omega$. The following properties are equivalent:*

   i)   *f has a graph defined in S1S;*
   ii)   *f is realized by an unambiguous Büchi 1-transducer.*

Remark 3.3. The unambiguous Büchi 1-transducers realize the synchronous functions.

Remark 3.4. For a S1S specification, to be a function can be expressed by a S1S formula so it is a decidable property.

Proposition 3.5. *Let be f a function whose graph is defined by a S1S formula. f is continuous iff f is realized by a sequential transducer with bounded delay.*

Remark 3.6. If f is continuous its graph is closed so it is a $\Pi_1^0(Auto)$ and we use a determinisation over the input to prove the result. This fact has been mentioned by [23] and [14]. In the opposite sens this is a particular case of a result of Frougny and Sakarovitch [11].

Example 3.1. A continuous function f is defined by the unambiguous Büchi 1-transducer (fig.14). The sequential transducer with bounded delay (fig.12) realizes f.

Remark 3.7. This function can not be realized by a 1-sequential transducer. This can be proved using the Büchi Landweber theorem on determination of finite state games – player I has a 1-sequential winning strategy so player II does not have a strategy. This is an example where Büchi says there is not synthesis for the specification. But in fact we can synthesize the specification

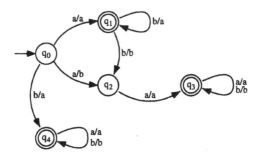

Figure 14

by a sequential circuit which outputs words. We think it will be interesting to find real examples of this kind for applications in non-terminating reactive finite state programs.

Proposition 3.8. *Let be f a function whose graph is defined by a S1S formula. The set of points of continuity of f is a $\Pi_2^0(Auto)$.*

Proof: It is easy to see that the set of points of discontinuity of f is defined by a S1S formula so by theo. 2.19 it is in *Auto*, by theorem 2.17 and theorem 2.3 it is $\Pi_2^0(Auto)$.   □

This proposition could be used if we are interested in the synthesis over a given domain.

Theorem 3.9. *Let be f a function realized by an unambiguous Büchi 1-transducer $\mathcal{T}$ (fig 15). f is not a continuous function iff there exists a pair of states (q_1, q_2) and a pair of words (u, v) such that*

> i)   *the loop $q_1 \xrightarrow{v} q_1$ does not meet a final state;*
>   *the loop $q_2 \xrightarrow{v} q_2$ meets a final state;*
>
> ii) $\sigma(q_0, u, q_1) = u_1;$
> $\sigma(q_0, u, q_2) = u_2;$ (1)
> $\sigma(q_1, v, q_1) = v_1;$
> $\sigma(q_2, v, q_2) = v_2;$ *and*
> $u_1 \neq u_2$ *or* $v_1 \neq v_2.$

Proof: Let there be a pair of states q_1, q_2 and a pair of words u, v such that condition (1) holds. Let the sequence of words be $(uv^n wt^\omega)$. We have

$$f(uv^n wt^\omega) = u_1 v_1^n w_1 t_1^\omega \text{ and } \lim_{n \to \infty} uv^n wt^\omega = uv^\omega,$$

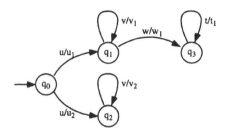

Figure 15

$$\lim_{n \to \infty} u_1 v_1^n w_1 t_1^\omega = u_1 v_1^\omega, \text{ but } f(uv^\omega) = u_2 v_2^\omega \neq u_1 v_1^\omega :$$

f can not be continuous because the limit of the image of the sequence is not equal to the image of the limit of the sequence.

On the opposite sens the proof uses the characterization of Landweber on Muller automata to prove the assertion.   □

Lemma 3.10. *Let $\mathcal{T}$ be an unambiguous Büchi 1-transducer realizing a function f. For a pair of states (q_1, q_2), if there exists a pair of words (u, v) such that (1) holds then there exists a pair (u, v) with $|uv| \leq 2n^2$ such that (1) holds.*

Corollary 3.11. *Let f be a function realized by an unambiguous Büchi 1-transducer. It is decidable whether f is not continuous.*

Remark 3.12. We have noted that for S1S specification to be a closed set is decidable (remark 2.18), and for S1S functional specification to be continuous can be expressed by a S1S formula. This gives two other ways to prove the corollary. But we prefer a combinatorial criterion which is visual. Our algorithm to prove non-continuity has a NP-complexity.

4 ASYNCHRONOUS FUNCTIONS

If there exists no transition $p \xrightarrow{u} q$ we denote the output by $\sigma(p, u, q) = 0$.

Definition 4.1. Let be $\mathcal{T} = \langle X, Y, Q, q_0, E, F \rangle$ an unambiguous Büchi transducer. Two states $q_1, q_2 \in Q$ (fig.16) are twinned iff $\forall u, v \in X^*$ the following condition holds

$$\left.\begin{array}{l} \sigma(q_0, u, q_1) = u_1 \neq 0 \\ \sigma(q_0, u, q_2) = u_2 \neq 0 \\ \sigma(q_1, v, q_1) = v_1 \neq 0 \\ \sigma(q_2, v, q_2) = v_2 \neq 0 \end{array}\right\} \Rightarrow u_1 v_1 u_1^{-1} = u_2 v_2 u_2^{-1}. \tag{2}$$

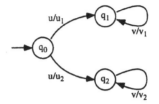

Figure 16

Definition 4.2. Let $\mathcal{T} = \langle X, Y, Q, q_0, E, F \rangle$ be an unambiguous Büchi transducer. $\mathcal{T}$ has the twinning property if any two states are twinned.

Remark 4.3. This property of transducers has been used to characterize the subsequential functions as a class of rational functions over finite words [7], [2].

Proposition 4.4. *Let* $u_1 u_2 v_1 v_2 \in Y^*$ [7]. *Then*

$$u_1 v_1 u_1^{-1} = u_2 v_2 u_2^{-1}$$

iff one the following conditions is verified

$\quad$ *i)* $\quad v_1 = v_2 = \varepsilon;$
$\quad$ *ii)* $\quad v_1 \neq \varepsilon \neq v_2$, *and there exists* $e \in Y^*$ *such that either*
$\quad$ *ii.a)* $\quad u_2 = u_1 e$ *and* $ev_2 = v_1 e;$
$\quad$ *ii.b)* $\quad u_1 = u_2 e$ *and* $ev_1 = v_2 e.$

Lemma 4.5. *Let be* $\mathcal{T}$ *an unambiguous Büchi transducer. For a pair of states* (q_1, q_2) *if there exists a pair of words* (u, v) *such that* (2) *does not hold then there exists a pair* (u, v) *with* $|uv| \leq 2n^2$ *such that* (2) *does not hold* [7].

Corollary 4.6. *Let be* $\mathcal{T}$ *an unambiguous Büchi transducer. It is decidable whether or not* $\mathcal{T}$ *has the twinning property* [7].

Example 4.1. The unambiguous Büchi transducer $\mathcal{T}$ (fig.17) which realizes the function $f : X^\omega \to X^\omega$ has the twinning property

$$\begin{cases} f(a^\omega) = a(ca)^\omega = (ac)^\omega; \\ f(a^{n+1}bX^\omega) = a(ca)^n bX^\omega, \ n \geq 0; \\ f(a^{n+1}cX^\omega) = (ac)^{n+1}cX^\omega, \ n \geq 0; \\ f(bX^\omega) = bX^\omega; \\ f(cX^\omega) = cX^\omega . \end{cases}$$

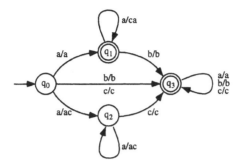

Figure 17

Theorem 4.7. *Let be f a sequential function. For each unambiguous Büchi transducer $\mathcal{T}$ realizing f, $\mathcal{T}$ has the twinning property.*

Example 4.2. Let be a function $f : X^\omega \to Y^\omega$ realized by the unambiguous Büchi transducer $\mathcal{T}$ (fig.18). f is continuous but f is not sequential.

$$\begin{cases} f(a^\omega) = (aa)^\omega = a^\omega; \\ f(a^n b X^\omega) = a^{2n} b X^\omega, \ n \geq 0; \\ f(a^n c X^\omega) = a^n c X^\omega, \ n \geq 0 . \end{cases}$$

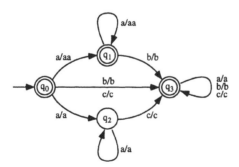

Figure 18

This function f is continuous. Let be the alphabets $A = \{a, b, c, x, y, z, t\}$, $B = \{a, b, c\}$, $C = \{a, b, c, aa\}$. Let us consider the set $K \subset A^\omega$ recognized by the automaton (fig.19).

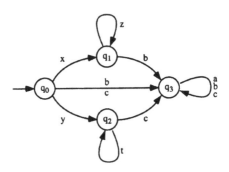

Figure 19

Let G be the graph defined by the unambiguous Büchi transducer $\mathcal{T}$. K is a $\Pi_1^0(Auto)$ so it is a compact set. Let us consider the following bimorphism (φ, ψ) defined by:

$$
\begin{array}{ll}
\varphi(a) = a; & \psi(a) = a; \\
\varphi(b) = b; & \psi(b) = b; \\
\varphi(c) = c; & \psi(c) = c; \\
\varphi(x) = a; & \psi(x) = aa; \\
\varphi(y) = a; & \psi(y) = a; \\
\varphi(z) = a; & \psi(z) = aa; \\
\varphi(t) = a; & \psi(t) = a.
\end{array}
$$

We have $(\varphi, \psi)(K) = G(f)$. (φ, ψ) is continuous so $(\varphi, \psi)(K)$ is compact and f is continuous since $G(f)$ is closed.

However f is not a sequential function because $\mathcal{T}$ does not have the twinning property:

$$
\left\{
\begin{array}{l}
u_1 = aa \\
u_2 = a \\
v_1 = aa \\
v_2 = a
\end{array}
\right.
\quad \text{so } (u_1 v_1 u_1^{-1} \neq u_2 v_2 u_2^{-1}).
$$

Example 4.3. The unambiguous Büchi transducer $\mathcal{T}$ (fig.17) has the twinning property. The function realized by $\mathcal{T}$ is also realized by the sequential transducer $\mathcal{T}'$ (fig.20).

Proposition 4.8. *Let there be a function* $f : X^\omega \to Y^\omega$ *realized by a unambiguous Büchi 1-transducer* $\mathcal{T}$. *We have*

f *is not continuous* $\Longrightarrow \mathcal{T}$ *does not have the twin property.*

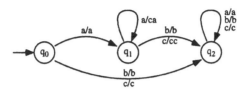

Figure 20

The continuous criteria (1) and the twinning criteria (2) are similar. For synchronous functions, (2) is a necessary and sufficient criteria to have sequentiality. For asynchronous functions we would extend this result to obtain the decidability of sequentiality (for rational functions on finite words this is a decidable property) [2].

Example 4.4. The unambiguous Büchi transducer $\mathcal{T}$ (fig.21) realizes the following function f:

$$f(\alpha) = \begin{cases} 0^\omega & \text{if } \alpha = 0^\omega, \\ 1^\omega & \text{if } \alpha \text{ contains at least one 1.} \end{cases}$$

Obviously, this function f is not continuous and the transducer $\mathcal{T}$ does not have the twinning property.

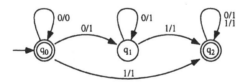

Figure 21

5 CONCLUSION

Our continuous function example (fig.18) is a real algorithm. It can be implemented with a stack. While we see a we write a and store a in a stack. When we see c we write c and after we just rewrite the input. But if we see b then while the stack is not empty we write the top of the stack and pop, then when the stack is empty we write b and after we just rewrite the input.

Continuous functions defined by unambiguous Büchi transducer are recursive functions. There are closed links with the synchronous decision diagram of Vuillemin [24]. As it was remarked by Klauss Winkelmann our continuous function example (fig.18) is an online function ($\forall \alpha, \beta \quad d(f(\alpha), f(\beta)) \leq d(\alpha, \beta)$). We hope our work will be used in the context of non-terminating reactive programs.

To finish, we note that the example (fig.18) proves that we can not extend the Büchi and Landweber theorem on finite state games (they said that if a player has a strategy then he has a 1-sequential strategy) to ω-rational relations of [11]. If we take for specification the graph of $S \subset (X \times Y)^\omega$, the player II has a unique winning strategy which is not a sequential function.

6   REFERENCES

[1] Arnold, A., "Rational languages are unambiguous", Theor. Comp. Sci. 26, 1983 pp. 221–223.

[2] Berstel, J., "Transductions and Context-Free Languages", Stuttgart, Teubner, 1979.

[3] Büchi, J.R., "On a decision method in restricted second order arithmetic", in Proc. Int. Congr. Logic, Method. and Philos. of Science (E. Nagel et al., eds.), Stanford Univ. Press, Stanford, 1962, pp. 1–11.

[4] Büchi, J.R., Landweber, L.H., "Solving sequential conditions by finite-state strategies", Trans. Amer. Math. Soc. 138, 1969, pp. 295–311.

[5] Burch, J.R., Clarke, E.M., Mc Millan, K.L., Dill, D.L., Hwang, L.J., "Symbolic model checking: 10^{20} states and beyond", Informations and Computation 98(2), 1992, pp. 142–170.

[6] Büttner, W., Winkelmann, K., "Equation solving over 2-adic integers and applications to the specification, verification and synthesis of finite state machines", 1995, to be published.

[7] Choffrut, C., "Une caractérisation des fonctions séquentielles et des fonctions sous-séquentielles en tant que relations rationnelles", Theoretical Computer Science 5, 1977, pp. 325–338.

[8] Church, A., "Logic, arithmetic and automata", Proc. Intern. Congr. Math. 1962, Almquist and Wiksells, Uppsala, 1963, pp. 21–35.

[9] Clarke, E., Grumberg, O., Long, D., "Verification tools for finite-state concurrent systems", A Decade of Concurrency (J.W. de Bakker et al., eds), Lecture Notes in Computer Science 803, Springer-Verlag, Berlin, 1994, pp. 124–175.

[10] Eilenberg, S., "Automata, Languages and Machines", vol A, Academic Press, New York, 1974.

[11] Frougny, C., Sakarovitch, J., "Rational relations with bounded delay", Rapport d'activité Laboratoire Informatique Théorique et Programmation 90.83, Institut Blaise Pascal, 1990.

[12] Ginsburg, S., Rose, G.F., "A Characterization of machine mappings", Canadian Journal of Mathematics 18, 1966, pp. 381–388.

[13] Landweber, L.H., "Decision problems for ω-automata", Math. Syst. Theory 3, 1969, pp. 376–384.

[14] Latteux, M., Timmerman, E., "Rational ω-transductions", Laboratoire d'Informatique fondamentale de Lille, Publication n IT 176 90, 1990.

[15] Mc Naughton, R., "Testing and generating infinite sequences by a finite automaton", Inform. and Control 9, 1966, pp. 521–530.

[16] Moschovakis, Y.N., "Descriptive set theory", North-Holland, 1980.

[17] Nivat, M., Perrin, D., "Automata on infinite words", Ecole de Printemps LNCS 192, 1984.

[18] Nökel, K., Winkelmann, K., "Controller synthesis and verification: a case study", C. Lewerentz, Th. Linder, Formal Development of Reactive Systems, Case Study Production Cell, Springer Lecture Notes in Computer Science, 891, Berlin, Heidelberg, 1995.

[19] Perrin, D., Pin, J.E., "Mots infinis", Laboratoire Informatique Théorique et Programmation, LITP 93.40, 1993.

[20] Staiger, L., "Sequential mappings of ω-languages", Math. Syst. Theory 3, 1987, pp. 376–384.

[21] Thomas, W., 1990, "Automata on infinite objects", Handbook of Theoretical Computer Science, Vol B, North-Holland, Amsterdam.

[22] Thomas, W., 1994, "On the synthesis of strategies in infinite games", Institut für Informatik und Praktische Matematik, Christian-Albrechts-Universität Kiel.

[23] Trakhtenbrot, B.A., Barzdin, Y.M., "Finite automata", North-Holland, Amsterdam, 1973.

[24] Vuillemin, J., "On circuits and numbers", IEEE Trans. on Computers 43:8:868–27,79, 1994.

[25] Wagner, K., "On ω-regular sets, Information and control 43, 1979, pp. 123–177.

9

AUTOMATIC GENERATION AND OPTIMISATION OF MARKOV MATRICES

A. Cabarbaye

CNES, 18 avenue Edouard Belin 31055 Toulouse France
Tel. 33 5 61 28 27 41 - Fax. 33 5 61 28 22 31
e-mail: Andre.Cabarbaye@cst.cnes.fr

ABSTRACT

This paper presents computer processing methods which automatically generate the Markov matrix of a system from logical expressions, to make easier the reliability and availability evaluations. These methods can take into account any relationships between transition rate values and configurations of the states of the system (cold redundancies, conditional maintenance..). Moreover, they allow the dimension of the matrix to be optimised by grouping the equivalent states.

1 INTRODUCTION

Markov processes can be used to model the behaviour of many systems and to calculate reliability/availability much more precisely than through simulations (Monte-Carlo). However, this type of model requires a complex construction for more than around ten states, and this cannot be done manually. Moreover, the number of states increases exponentially with the number of elements in the system (2^n for n elements with two states: correct operation and failure), often making calculations very difficult and limiting the size of the systems studied.

The aim of this article is to present computer processing methods which automatically generate the Markov matrix of a system, using logical expressions which characterise its operation, and which allow the dimension of the matrix to be optimised by grouping some of the states together.

2 LOGIC ANALYSER BASED ON THE PRINCIPLE OF INSERTION

The Markov matrix of a system of n elements can be constructed in a simple way using insertion. The combinations of the states "correct operation" (a) and "failure" (na) of the various elements (a, b, c...) are arranged as follows:

				1	2	3	4	5	6	7	8
c	b	a	1	-	λa	λb		λc			
c	b	na	2	μa	-		λb		λc		
c	nb	a	3	μb		-	λa			λc	
c	nb	na	4		μb	μa	-				λc
nc	b	a	5	μc				-	λa	λb	
nc	b	na	6		μc			μa	-		λb
nc	nb	a	7			μc		μb		-	λa
nc	nb	na	8				μc		μb	μa	-

3 element matrix

Coefficients λ_{ij} represent the transition rates from line states i (left) to column states j (top). Transition rates λ_a and μ_a are respectively the failure and repair rates of element a.

The states for which the system is available can be simply defined using a logical expression relating the various elements. Any relationships linking certain transition rate values to configurations of the states of the system can also be expressed in this way. This type of relationship is used for example for conditional maintenance or for cold redundancies ($\lambda^* = \lambda/10$ for an element which is switched off). The approach is illustrated in the following example:

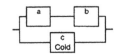

Available states : **(a and b) or c**

Dependency relationships : **a and b** $\Rightarrow \lambda c^*$

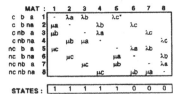

This method for automatically generating the Markov matrix is currently included in the Supercab software, which is used at CNES for some assessments. The states for which the system is available are defined by a logical expression of the form n(axb+n(c+exnf)) using the logical operators OR (+), AND (x), and NOT (n), where letters a, b, c, e, and f represent the elements of the system. Dependency relationships can be written using similar expressions. The software tool uses insertion to construct the system matrix. Dependent links are taken into account by modifying the transition rate values using the corresponding logical expressions. The user need not built the matrix itself, which means that it is relatively easy to determine the availability of a complex system. This is illustrated in the following example:

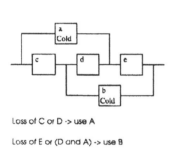

Loss of C or D -> use A

Loss of E or (D and A) -> use B

	Available states :	Logical expressions
	hr⁻¹	cxdxe+axe+cxb
λ_a	7,00E-05	
λ_a*	7,00E-06	cxd
μ_a	9,00E-03	
λ_b	8,00E-05	
λ_b*	8,00E-06	cx(d+a)
μ_b	7,00E-03	
λ_c	4,50E-05	
μ_c	8,00E-03	
λ_d	7,50E-05	
μ_d	6,00E-02	
λ_e	9,00E-04	
μ_e	9,50E-03	

Availability

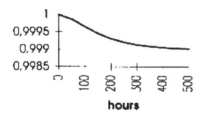

In spite of the attractiveness of this processing method, one major disadvantage is that the dimension of the matrix generated (2n) can rapidly restrict its use. A method for grouping states which minimises this problem is therefore presented below.

3   GROUPING STATES

Many groups of states are possible within a system as long as not all of the elements are considered to be individually repairable. Indeed, the configuration of the system often results in some states being equivalent. If there are symmetries in the system architecture, the number of equivalent states can be significantly increased.

The following example illustrates how states can be grouped together:

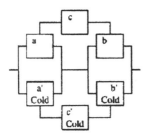

Hypotheses :

a and a' identical elements

b and b' identical elements

c and c' identical resources

The number of elements used simultaneously is minimised

Losing resource c (power supply of a and b) has the same effect on the system as losing both a and b. Using equivalences of this type and the set of symmetries which may be observed in the architecture allow the 64 (2^6) states of the system to be grouped into 6 distinct states.

The reduced Markov matrix then becomes:

	1	2	3	4	5	6
No failure : 1	-	$\lambda a + \lambda a^*$	$\lambda b + \lambda b^*$	0	$\lambda c + \lambda c^*$	0
Loss of a or a' : 2	0	-	0	λb	$\lambda b^* + \lambda c^*$	$\lambda a + \lambda c$
Loss of b or b' : 3	0	0	-	λa	$\lambda a^* + \lambda c^*$	$\lambda b + \lambda c$
Loss of a and b' or a' and b : 4	0	0	0	-	0	$\lambda a + \lambda b + 2\lambda c$
Loss of all redundancies : 5	0	0	0	0	-	$\lambda a + \lambda b + \lambda c$
Loss of the system : 6	0	0	0	0	0	0

Repair rates for entire blocks can then be introduced into the reduced matrix (e.g.: repair rate of block abc).

In many cases, this method for grouping states prevents the number of combinations of states from becoming too large. However, it requires an in-depth and often tedious analysis of how the system operates, and can lead to errors such as incorrect grouping (state 4 often forgotten in the above example). Because of this, a way of making the analysis automatic was developed.

4   AUTOMATIC METHOD FOR GROUPING STATES

A method for grouping states was designed and a model developed. First, both the logical expressions defining the system's available states and any dependency relationships, are decomposed into simple elements. Then the states which result in identical logical states for each of these simple elements are grouped together. The following example illustrates the approach:

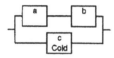

States	a×b	c	Grouped states
c b a	1	1	→ c b a
c b na	0	1	1
c nb a	0	1	1 → c nb na
c nb na	0	1	1
nc b a	1	0	→ nc b a
nc b na	0	0	1
nc nb a	0	0	1 → nc nb na
nc nb na	0	0	1

Available states: a×b+c
Dependency relationships : λc* if a×b

Two simple and distinct elements ab and c are obtained by decomposing the logical expressions. 1 represents the state "correct operation", 0 represents "failure". For identification purposes, each group of states takes the name of the state of the group which contains the most failed elements (c b na, c nb a, c nb na ⟶ c nb na). This method of preliminary grouping is very effective for many real situations. The examples analysed above give the following results:

5   CONCLUSION

Assessing reliability/availability can enable both the system architecture (redundancies, cross-strapping, reconfiguration procedure, etc.) and maintenance policies (spare parts batch, preventive maintenance, availability of technical

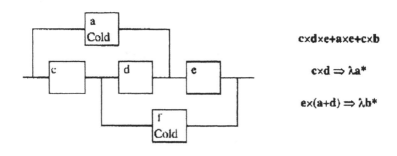

cxdxe+axe+cxb

cxd ⇒ λa*

ex(a+d) ⇒ λb*

Grouped states	1	2	3	4	5	6	7	8
a b c d e: 1	-	λa*	λb*	λd		λc	λe	
na b c d e: 2		-			λb*		λd+λe	λc
a nb c d e: 3			-		λa*	λd+λc		λe
a b c nd e: 4				-		λb*+λc	λa+λe	
na nb c d e: 5					-			λc+λd+λe
a nb nc nd e: 6						-		λa+λe
na b c nd ne: 7							-	λb+λc
na nb nc nd ne: 8								-

The dimension of the resulting matrix is 8 instead of 32 (2^5).

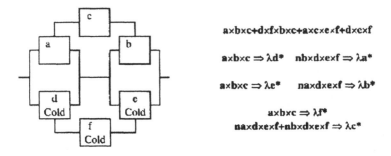

axbxc+dxfxbxc+axcxexf+dxexf

axbxc ⇒ λd*   nbxdxexf ⇒ λa*

axbxc ⇒ λe*   naxdxexf ⇒ λb*

axbxc ⇒ λf*
naxdxexf+nbxdxexf ⇒ λc*

Grouped states	1	2	3	4	5	6	7	8	9	10
a b c d e f : 1	-	λa	λb	λd^*	λe^*			λc	λf^*	
na b c d e f : 2		-				λe		$\lambda b^*+\lambda c^*$		$\lambda d+\lambda f$
a nb c d e f : 3			-				λd		$\lambda u^*+\lambda c^*$	$\lambda c+\lambda f$
a b c nd e f : 4				-			λh		$\lambda e^*+\lambda f^*$	$\lambda a+\lambda c$
a b c d ne f : 5					-	λa			$\lambda d^*+\lambda f^*$	$\lambda b+\lambda c$
na b c d ne f : 6						-				$\lambda b+\lambda c+\lambda d+\lambda f$
a nb c nd e f : 7							-			$\lambda a+\lambda c+\lambda e+\lambda f$
na nb nc d e f : 8								-		$\lambda d+\lambda e+\lambda f$
a b c nd ne nf : 9									-	$\lambda a+\lambda b+\lambda c$
na nb nc nd ne nf :10										-

The dimension of the resulting matrix is 10 instead of 64 (2^6).

staff, etc.) to be optimised. As a result, the global cost of a product can often be significantly reduced. However this assessment is sometimes difficult to carry out, particularly for non-specialist designers.

The work described here helps solve this problem by providing users with tools which only require knowledge of the logical operation of the system concerned. Moreover, the suggested method for grouping states should allow Markov techniques to be extended to complex systems with many interdependent elements.

Further research could improve this method in order to identify and exploit symmetries in the architecture. Indeed, for most applications, the system contains identical elements which can lead to further grouping of states.

6 REFERENCES

[1] A. Pages, M. Gondran, Fiabilité des systèmes, Edition Eyrolles, Paris 1980.

[2] A. Villemeur, Sûreté de fonctionnement des systèmes industriels, Edition Eyrolles, Paris 1987.

[3] E. Lourme, Modélisation d'un système par matrice de Markov, Sup Aéro Toulouse 1995.

[4] J. Devooght & B. Tombuyses, Aggregation methods for reliability and availability calculations, Elsevier Science Limited, Oxford UK, 1996.

10

FAULT MODELING IN SPACE-BORNE RECONFIGURABLE MICROELECTRONIC SYSTEMS

A. Rucinski*,
N. Valverde*,
C. Baron**,
P. Bisgambiglia***,
D. Federici***,
J-F. Santucci***

University of New Hampshire, U.S.A.
*** INSAT/DGEI/GERII, Toulouse, France*
**** University of Corsica, France*

ABSTRACT

With diminishing budgets of space agencies around the World, it becomes critical to search for lower cost solutions to keep exploration programs alive. One of the cost contributing factors is the common use of rad-hard electronic devices for space applications. This paper questions this requirement by proposing an alternative which employs highly adaptable microelectronic architecture potentially absorbing the impact of space born defects. The FPGA soft programmable device under study is driven by a rad hard 80186 microprocessor. The proposed experiment, called TRIAD, enables the validation of different fault models in space borne systems, with the expectation of behavioral fault models being most attractive.

1 INTRODUCTION

Emerging paradigms of *Reconfigurable Systems* [1], [2] which include RAMs, FPGAs [3], PCMCIA cards, and hard disks (well established technologies) on one hand and *Microarchitectures* (an emerging low end category of microelectronic systems) on the other, require addressing the dependability issue in a novel, coherent, and integrated fashion. The testing issue addressed in this paper is considered in the context of applying reconfigurable systems in space which poses two additional challenges:

- The first one is the problem of *determining fault models in space-borne systems.* Today's radiation-hardened solutions, typically used in space applications, lag far behind the capabilities of mainstream microelectronic systems [4]. Shrinking federal budgets may further widen this gap, and even eliminate the rad-hard market. To expand the number of options for space-borne applications, this paper introduces a concept enabling semiconductor engineers to study rad-hardness alternatives using an approach based on a TRIAD: highly dependable systems, evolving technologies, and advanced packaging. The focus is on *Behavioral Faults* [5], [6] used as metrics to characterize the technology-dependability impact. Better understanding of physical defects manifested at a lower than the system level may ease the need for radiation hard solution in space by employing massive redundant options in reconfigurable systems.

- The second issue is connected with the development of *small, light*, and even *nanosatellite technologies* [7] in both scientific and commercial missions. Rad hard technology is generally expensive, but especially unacceptable in a COTS oriented small satellite technology. Rad hard approach is by definition counterproductive since it uses "frozen" architectural and systems developments. In addition, radiation effects are typically considered at the system level (e.g. the maximal radiation dosage is proportional to the Mean Time To Failure (MTTF) which implies a *system level reliability model* [8]. This model does not take into account the logical and functional structure of a microelectronic device). Thus, new models at a lower level of abstraction need to be derived.

The TRIAD experiment can be conducted either on Earth or in Space, or both, with the following trade-offs in mind:

- An Earth experiment is very expensive with a single test typically free of charge. In Space the cost can be very reasonable since it is sufficient to conduct TRIAD as a secondary space experiment with a low priority communication requirements.

- Artificial radiation environment created in a laboratory may not be sufficiently accurate to conduct this experiment.

In the next section, we present the TRIAD architecture and the scenario of Fault Model Validation. In the section three, VHDL description of devices under test (a non-redundant register and a triple redundant register with voter).

2   THE TRIAD ARCHITECTURE

The TRIAD architecture is a test bed for conducting trade-off studies between technologies, fault-tolerance, and testing in space:

. **Architecture:**   Figure 1 represents a block-level diagram of the architecture system concept. The upper part of the design, an FPGA-based bus interface and 80186-based rad-hard controller, is implemented in technologies with high survivability in space. The lower part corresponds to the Unit Under Test (UUT), with uncertain space survivability. UUT represents a volume in three dimensional space determined by testability, technology, and fault tolerance selected for the experiment.

. **Dynamically Reconfigurable** UUT: An attractive solution for UUT appears is a dynamically reconfigurable architecture [1], [2]. In this case, a system using preprogrammed information about its desired configuration respond to outside environment by generating the "best" architecture. A good example is a soft-reconfigurable FPGA arrangement when a silicon device can change its configuration [9]. A merge of the concept and technology has lead to "reconfigurable computing" [10] as well as to "morphological" systems [11]. Test and diagnosis in FPGAs using this approach is described in [12] where physical defects are diagnosed through a variety of logical architectures reconfigured for easy testing. We are proposing a similar, but modified approach since neither the adequate architecture nor the optimal fault model suitable for space borne system is determined so far. Thus, reconfiguration enables us to use a microelectronic system which gradually achieves desired optimum through its flexible architecture.

. **The Scenario of Fault Model Validation:**   The flow diagram for the TRIAD experiment is depicted concisely in Figure 2. Each level of consideration reflects the flexibility and expandability of the approach. For example, the technology level assumes a repetitive TRIAD experiment for a variety of technologies. There is at least a pair of identical microprocessor based systems: one aboard the satellite (the SPACE copy) and another on Earth (the EARTH copy). Only the EARTH version is equipped with a simulation/synthesis CAD

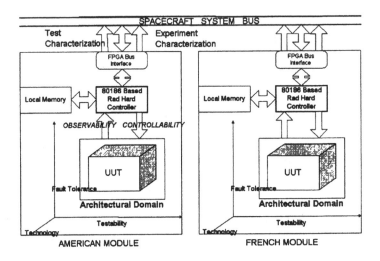

Figure 1   The TRIAD Concept

package. For a given technology and a given level of abstraction the corresponding *fault model* and the logical architecture described in VHDL is considered. The VHDL model is synthesized and the target physical configuration is up-linked to the satellite. As a result, both systems are reconfigured the same way, followed by simultaneous testing of both copies with the EARTH copy treated as a *golden unit*. The discrepancy in the system recognitions leads to *fault detection* followed by a *diagnosis phase*. A *non-empty fault map* employs a reconfiguration scheme accomplished again by the two versions with a proper synthesis and upload. The procedure is repeated until a reconfiguration scheme absorbs the effect of a fault resulting in the validation of the fault model. Exceptions, i.e. when absorption is impossible, are handled either by a different reconfiguration or a different fault model [13].

. Logical versus Physical Approach: Two approaches are considered: the logical approach when a physical location of logic under study is not relevant and the physical approach in an opposite case. After *detecting a defect*, the logical approach gradually increases the granulity of logic until the effect of defect is masked off resulting in the fault model validation. Multiple and single defects are handled likewise. After *diagnosing a defect*, the physical approach eliminates a block which contains the defect and resynthesizes the structure. Multiple defects may be difficult to handle in the physical approach. However, this approach suggests a new diagnostic method of "roving diagnosis" which

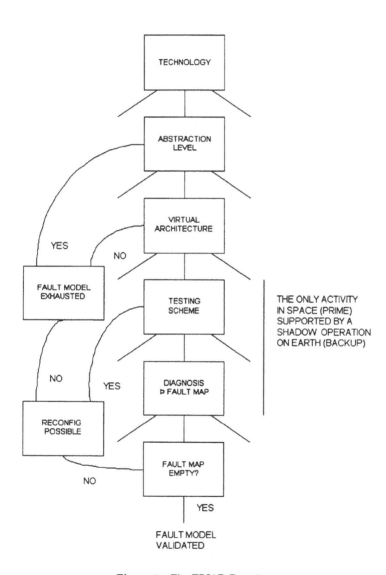

Figure 2 The TRIAD Experiment

seeks for malfunctioning blocks one at a time. A question can be raised whether elimination of or at least reduced diagnosis is feasible by using "reverse engineering" [14] or signature based information. For example, XILINX FPGA have some readback capabilities useful in this case.

. **Additional Assumptions:** The initial architecture is assumed to have structure V1 (developed on Earth and uplinked after synthesis to a satellite) which results in physical structure P1. The assumed fault model is based on [6] and testing is performed at the VHDL level (test vectors and test procedures are also developed on Earth and uplinked to the satellite) followed by testing in space. Both detection and diagnosis is performed at the VHDL level. Reconfiguration forced by malfunction is again performed on Earth and the results of synthesis (VHDL model V2 and physical structure P2 respectively) are uplinked and the process is repeated. Note, that it is assumed that V2 > V1 (meaning that a new fault tolerant configuration may absorb the effect of a fault detected in V1) as well as that P2 > P1 (a new more powerful configuration should result in a more robust, area demanding physical structure which eliminate the effect of a failure present in P1). It is envisioned that V1 structure contains a number of basic modules in VHDL which will "cover" the whole silicon area provided by a FPGA chip. It is also envisioned that V2 structure generates the maximum capacity physical implementation of modules after synthesis which a FPGA device can accommodate.

3   IMPLEMENTATION EXAMPLES

It is assumed that a microprocessor based system is based on the Intel 80186 rad-hard device with a configuration analogous in one being developed for the CATSAT project. UUT are soft programmable FPGA devices programmed using logic synthesis approach. The two VHDL modules (non-redundant and TMR based) of the device under study are as follows:

```
------------------------------------------------------------------
--
--File name: example1.vhd
--Title    : Register_ex
--Library  : synth
--Purpose  : Behavior description of a register_ex which is used for
--a reconfigurable system concept devoted for space applications
--IO       :
--    PORT <name>      <mode> <type>            <purpose>
--           DI      : IN   vlbit_1d(1 TO 8) ; data in
--           STRB    : IN   vlbit ;             transfert control
--           DS1     : IN   vlbit ;             enable control
--           NDS2    : IN   vlbit ;             enable control
--           DO      : OUT  vlvlbit_1d(1 TO 8) data out
------------------------------------------------------------------


Entity register_ex IS
  PORT (DI               : IN vlbit_1d(1 TO 8) ;
        STRB, DS1, NDS2 : IN vlbit ;
        DO               : OUT vlbit_1d(1 TO 8)
        ) ;
END register_ex;

ARCHITECTURE behavior OF register_ex IS
  SIGNAL reg           : vlbit_1d(1 TO 8);
  SIGNAL enbld         : vlbit ;
BEGIN --  behavior
  strobe: PROCESS (STRB)

  BEGIN --  strobe
    IF (STRB='1') THEN
      reg <= DI ;
    END IF;
  END PROCESS strobe;
  enable: PROCESS (DS1,NDS2)

  BEGIN --  enable
    enbld <= DS1 AND NOT(NDS2) ;
  END PROCESS enable;

  output: PROCESS (reg,enbld)
```

```
  BEGIN --  output
    IF (enbld='1') THEN
      DO <= reg ;
    ELSE
      DO <= X"11";
    END IF;
  END PROCESS output;
END behavior;  -- of register_ex
```

Figure 3: The FPGA Device Behavior

```
---------------------------------------------
--
-- File name : triple.vhdl
-- Title     : triple redundant register with voter
-- Library   : synth
-- Purpose   : example for reconfigurable FPGA for space application
-- IO        :
--
--    PORT <name>  <mode>  <type>              <purpose>
--           DI    : IN    vlbit_1d(1 TO 8) ; data in
--           STRB  : IN    vlbit ;            transfert control
--           DS1   : IN    vlbit ;            enable control
--           NDS2  : IN    vlbit ;            enable control
--           DO    : OUT   vlvlbit_1d(1 TO 8) data out
--
--
---------------------------------------------

ENTITY triple IS
  PORT (
        DI                 : IN vlbit_1d(1 TO 8) ;
        STRB, DS1, NDS2 : IN vlbit ;
        DO                 : OUT vlbit_1d(1 TO 8)
        ) ;
END triple;
```

```
ARCHITECTURE behv OF triple IS
  COMPONENT register_ex
    PORT (
          DI                 : IN vlbit_1d(1 TO 8) ;
          STRB, DS1, NDS2 : IN vlbit ;
          DO                 : OUT vlbit_1d(1 TO 8)
          );
  END COMPONENT;

  COMPONENT voter
    PORT (
          DO1                : IN vlbit_1d(1 TO 8) ;
          DO2                : IN vlbit_1d(1 TO 8) ;
          DO3                : IN vlbit_1d(1 TO 8) ;
          DO                 : OUT vlbit_1d(1 TO 8)
          );
  END COMPONENT;

  -----------------------------------------------------
  --  signal declaration
  -----------------------------------------------------
  SIGNAL DO1  : vlbit_1d(1 TO 8);
  SIGNAL DO2  : vlbit_1d(1 TO 8);
  SIGNAL DO3  : vlbit_1d(1 TO 8);

BEGIN --  behv
  c100: register_ex PORT MAP (
                              DI          => DI,
                              STRB        => STRB,
                              DS1         => DS1,
                              NDS2        => NDS2,
                              DO          => DO1
                              );
  c200: register_ex PORT MAP (
                              DI          => DI,
                              STRB        => STRB,
                              DS1         => DS1,
                              NDS2        => NDS2,
                              DO          => DO2
                              );
  c300: register_ex PORT MAP (
                              DI          => DI,
```

```
                              STRB        => STRB,
                              DS1         => DS1,
                              NDS2        => NDS2,
                              DO          => DO3
                              );
        c400: voter PORT MAP (
                         DO1               => DO1,
                         DO2               => DO2,
                         DO3               => DO3,
                         DO                => DO
                         ) ;
END behv;  -- of triple
```

4 CURRENT AND FUTURE WORK

Our current work deals with several aspects :
i) The design of the devices under test using logic synthesis approach,
ii) Behavioural fault modeling and fault simulation,
iii) Implementation of the scenario of fault model validation.
We envision launching the TRIAD experiment aboard a satellite as part of a future international micro satellite designed and built by international teams of students.

5 REFERENCES

[1] A.Rucinski and J.L.Pokoski, "Polystructural, Reconfigurable, and Fault-Tolerant Computers", The IEEE 6th Int. Conf. on Distributed Computing Systems, Cambridge, Massachusetts, May 1986.

[2] J.P. Gray and T.A. Kean, "Configurable Hardware: A New Paradigm for Computation", Proc. Advanced Research in VLSI, ed. C.L. Seitz, March 1989.

[3] http://www.xilinx.com/programs/reconfig.htm

[4] A. McAuliffe, "Staying Alive: Rad-Hard ICs Continue to Populate Satellites", Military&Aerospace Electronics, vol. 7, No. 3, March 1996.

[5] J.F. Santucci, G. Dray, N. Giambiasi, M. Boumedine, "Methodology to Reduce Computational Cost of Behavioral Test Pattern Generation using Testability Measures", 29th IEEE/ACM Design Automation Conference, 1992.

[6] J.F. Santucci, A.L. Courbis and N. Giambiasi, "Behavioral Testing of Digital Circuits", Journal of Microelectronic Systems Integration, March 1993.

[7] K. Levenson and K. Reister, "A High Capability, Low Cost University Satellite for Astrophysical Research", 8th Annual AIAA/USU Conference on Small Satellites, Logan UT, August/September 1994.

[8] K. LaBel, "Radiation Effects and Analysis Home Page", WEB Site, http://flick.gsfc.nasa.gov/radhome.htm/

[9] The Programmable Logic Data Book., Xilinx, Inc., 1993.

[10] R.T. Maniwa, "Reconfigurability: logical computing", Integrated System Design, June 1995.

[11] Tien-Hsin Chao, "Dynamically reconfigurable optical morphological processor", NASA Tech Briefs, vol. 20, No. 2, February 1996.

[12] Ch. Stroud et al., "Built-in self-test of logic blocks in FPGAs", submitted to the 14th IEEE VLSI Test Symposium, Princeton, NJ, April/May 1996.

[13] J.A. Clark and D.K. Pradhan, "Fault injection: a method for validating computer-system dependability", Computer, June 1995.

[14] C. Baron and J.C. Geffroy, "Synthesis of Identification Methods: Application to Performance Testing", Journal of Microelectronic Systems Integration, December 1994.

PREVENTION OF REPLICATION INDUCED FAILURES IN THE CONTEXT OF INTEGRATED MODULAR AVIONICS

P. D.V. van der Stok*,
P. T.A. Thijssen**

*Dept. of Computing Science, Eindhoven University of Technology,
P.O. Box 513, 5600 MB Eindhoven

**Philips Sound & Vision, BG Media Systems
P.O. Box 80002, 5600 JB Eindhoven
Netherlands

ABSTRACT

The motivation for Integrated Modular Avionics (IMA) is presented. The required high availability and improved maintenance efficiency dictate requirements on the consistency of data used by replicated software components. It is shown that a reliable multicast facility is needed to fulfill the consistency requirement.

Propagation of failures should be prevented. An additional consistency requirement states that software components should consider the same resources as failed at the same time. It is shown how a membership algorithm can satisfy this requirement. The time bounds on communication and failure detection propagation are calculated.

1 INTRODUCTION

An avionic application is a good example of an embedded system. Two parts can be discerned the system under control (the airplane) and the controlling system (the set of interconnected processors with software). The state of the controlled system is determined by the state of the controlling system and vice versa. These systems are notoriously complex to build because the system's response not only depends on the invoked function and its input parameters but also on the controlling system's state and its history. It is clear that rigid guidelines are needed to design such systems, especially when the failing of the system has important consequences in financial or social terms.

This is especially true for Avionics concerned with the control of the airplane during its flight from its point of departure until its destination. Failures of the controlling system can lead to loss of money, loss of aircraft or parts of it and loss of life.

During the flight of an aircraft, different phases can be discerned like: taxiing, takeoff, landing 1, landing 2, level flight, ... During flight phases, components have criticality levels that describe the seriousness of the consequences of their failure. A level is associated with its maximal probability of failure during one hour of flight. Five different criticality levels are discerned (with associated Failure probability): Catastrophic (10^{-9}), Hazardous (10^{-7}), Major (10^{-5}), Minor (10^{-3}) and No effect(-). It is clear that the more probable failure of a component with a low criticality level should in no way jeopardize the correct execution of the other components. Therefore, avionic systems were constructed such that components of a given criticality level were supported by dedicated hardware with no electrical or electronic connections between components of different criticality level.

The controlling system not only needs to meet the very demanding reliability and functional specifications, also the purchasing and maintenance costs should be as low as possible. An important contribution to the maintenance costs is caused by the large number of types of equipment. A large and expensive stock of any type of equipment must be provided by an aircraft operator to keep the number of flight delays within acceptable limits.

The driving force behind the introduction of Integrated Modular Avionics (IMA) is (1) lower maintenance cost, (2) higher availability and reliability, and (3) the need for more sophisticated and fuel saving control of equipment.

Ad 1) IMA provides a set of electrical, electronic, mechanical and software standards in which manufacturers can produce a standard set of modules with different software functionalities that can be used for a wide range of aircraft types from different manufacturers. The aim is to provide a stock of standard hardware modules with a reduced set of types that can be configured to the required functionality. Such a stock composed of a limited number of module types leads to reduced maintenance costs. The application of on-board surveillance and tests reduces the number of replacements of correctly functioning modules.

Ad 2) IMA provides a set of interconnected processors that can be configured to execute a number of required functions. Functions can be replicated over different processors to meet the specified availability and reliability criteria. The number of possible configurations that is larger for IMA than for non-IMA systems allows the meeting of reliability and availability requirements with a degraded system for lower costs than is possible with the non-IMA systems.

Ad 3) The modular composition of an IMA system makes it amenable to extensions and growth. The growing need for computing power to fly an aircraft with lower costs within ever smaller operational margins makes it attractive to have systems that can be upgraded during the lifetime of the aircraft without major modifications to all not directly implicated components. The IMA concept provides such a framework.

The communication needs between the IMA modules and the possibilities to reconfigure the IMA system necessitate the introduction of hardware and software components that are shared by functions of different criticality level. This is the most important difference from a reliability point of view of the IMA system with respect to the more traditional systems. The sharing of hardware components has an impact on the reliability calculation of the total controlling system. Software dependencies should not enhance these dependencies and care should be taken that no hidden dependencies are introduced. The prevention of hidden dependencies associated with replication and a shared communication medium are the subject of this paper.

Some IMA concepts and communication standards are introduced in section 2. A software architecture that supports failure separation is proposed in section 3. It is shown how the introduction of a membership algorithm [11] prevents the introduction of hidden dependencies in section 4.

2 INTEGRATED MODULAR AVIONICS

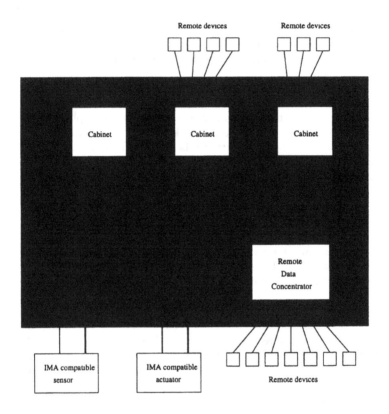

Figure 1   IMA with its environment

In Fig. 1, the shaded area represents the controlling system (IMA Host System). The devices, resources and actuators are part of the controlled system (the aircraft). An IMA Host System is composed of a number of Remote Data Concentrators (RDC) and cabinets interconnected by an airplane bus [2]. The currently proposed airplane bus standard is ARINC629 [1] described in section 2.1. A cabinet houses Line Replaceable Modules (LRM) electrically interconnected by the cabinet backplane. The mechanical and electrical standard is described in section 2.2 according to ARINC document [3]. Standard LRM's are core modules and gateway modules. Core modules contain a processor. Processors placed in different cabinets communicate over the airplane bus via the gateways. For reliability reasons gateways are replicated. Processors have local clocks which are assumed to be synchronized. For a processor

p a clock function $C_p(t)$ is defined that returns the value of the local clock at physical time t. For any two correct processors p and q and given constant ϵ, it is assumed that $\mid C_p(t) - C_q(t) \mid < \epsilon$. Assuming that in short intervals clocks can be approximated with: $d\,C_p(t)/dt \approx 1$, the following relation holds: $C_p(t_1) = C_q(t_2) \Rightarrow \mid t_1 - t_2 \mid < \epsilon$.

A core module executes partitions that execute in isolation from other partitions on the same core module. *Private* memory is accessible to only one partition and *shared* memory is accessible to all partitions of a core module.

The avionic application is subdivided in avionic functions. An avionic function acts on the equipment of the aircraft based on sensor input from the aircraft environment to orient the aircraft in space and time in accordance with the aircraft wanted behaviour. Examples of avionic functions are yaw-damping, auto-pilot, lift augmentation and others. An avionic function is implemented with an avionic system that represents the hardware configuration needed by the avionic function to meet its functional and availability requirements. An avionic system is a subset of the IMA Host System. Avionic systems may contain the same equipment e.g. the airplane bus.

An avionic function is realized by writing a software module that can be instantiated as one or more collaborating partitions. Several software modules can be written to satisfy the avionic function specification. These modules can use different avionic resources to meet the avionic function specifications. Due to high reliability, availability and hours of working after first failure, catastrophic avionic functions need to be replicated over more than one cabinet. Replication is handled in two ways: (1) installing *versions* or (2) installing *replicas*. Two partitions are called versions iff they realize the same avionic function but with different software modules. Two partitions are called replicas iff they realize the same avionic function with copies of the same software module. An avionic functions that is distributed over a set of core modules is realized by one partition per core module.

2.1 ARINC629

The ARINC specification 629 [1] defines a digital communication system in which *terminals* (RDC and cabinets) may transmit and receive digital data using a standard protocol communicated over electrically conducting or fiber optic media. The ability to transmit both periodic and aperiodic data in a bidirectional way is basic to the design. Each terminal is programmed to send data during globally defined intervals. Two protocols, the Basic Protocol (BP) and the Combined mode Protocol (CP), support point-to-point and broadcast protocols.

Basic Protocol

Three intervals are defined by the protocol:

- Transmit Interval (TI), common to all terminals,
- Sync Gap (SG), a quiet interval common to all terminals,
- Terminal Gap (TG), a quiet interval unique to each terminal.

Each individual terminal starts TI when it starts transmitting during an interval smaller than TI. A terminal may start transmitting when an interval TI has elapsed, no terminal has sent data during interval SG and during its personal interval TG no data is sent. The periodicity of the bus is determined by the load and the definition of the three intervals. All terminals are synchronized with each other via interval SG. Terminals do not interfere with each other due to two mechanisms dependent on the mode. In the aperiodic mode, a terminal starts when SG has elapsed. Because SG is attributed uniquely, only one terminal will start sending at a given time. In the periodic mode, a terminal cannot start sending before TI has elapsed and TI is chosen longer than the total expected transmission times of all terminals within a given period: TI > MFT. Minimum Frame Time (MFT) is defined as the sum of the bus occupation intervals of the terminals.

Correct functioning of protocols is guaranteed by timers. Timers are never completely synchronized. In the periodic mode this has as consequence that the order of sending within a given period is not always the same.

Dependent on the load and the definition of TI, all terminals either execute the periodic or the aperiodic protocol. Protocols cannot be combined.

An example of a periodic timing diagram is shown in Fig. 2.

Combined Mode Protocol

The CP protocol uses the same three intervals defined above. Additionally three levels are defined in which periodic and aperiodic messages can be combined within one period:

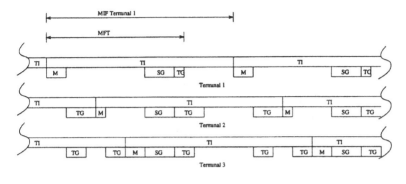

Figure 2   Possible timing diagram for ARINC 629 protocol

- level 1: periodic messages

- level 2: infrequent high priority messages (short duration)

- level 3: low priority aperiodic messages

During a period a terminal may send only one level1 and level 2 message. Level 3 messages can have an unspecified duration and as many as needed can be sent. The duration of level 1,2 and 3 messages within one period is bounded.

2.2 ARINC659

A replicated bus based on the ARINC659 standard [3] provides the communication between LRM's in a cabinet. Transmission and reception to and from the bus are table driven. The tables of all LRM's in one cabinet need to be the same for consistency reasons. Bus time is divided in a series of windows, each window containing a single message. Data is transferred according to a predetermined transfer schedule. Tables define the length of each window and the transmitter(s) and receiver(s) within this window. Transfer is guaranteed under several failure conditions. The bus transfer schedule is organized in cyclic

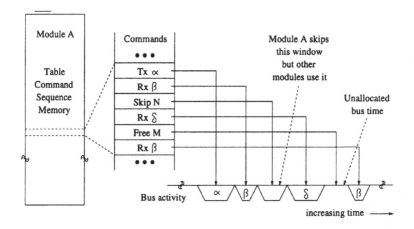

Figure 3   Possible table organization for ARINC 659 protocol

loops of constant length set by the sum of the individual window lengths. A possible table setting for a given LRM is shown in Fig. 3.

Each window has either a unique transmitter or a limited set of candidate transmitters obeying the so called Master/Shadow protocol. When the master starts sending at the allocated time, no other shadow transmitter transmits. If the master and some shadows fail, the first correct scheduled shadow transmits a message, all later scheduled shadows do not transmit.

Windows are organized in cyclic frames. More than one frame, possibly corresponding with different flight phases, is programmed. Special control messages are foreseen to change from a given frame to another one in a controlled manner for all LRM's in cabinet.

3   IMA SOFTWARE ARCHITECTURE

An avionic function needs two types of material resources: (1) *control resources*: control system hardware needed for its execution (shaded area in Fig. 1) and (2) *avionic resources*: actuators and sensors that are needed by the avionic functions to establish the position and movement of the aircraft (non shaded area in Fig. 1). The results produced by a partition can serve as input to another partition. When functions are replicated over versions or replicas, the results of correct partitions should be identical or within a given range. When results of resources or partitions are wrong, it is important to detect this and ascertain its cause for later maintenance. When the failures of resources or partitions are detected, a partition can decide to ignore the associated results. When a partition ignores a result, the other partitions using the same results should equally ignore them to arrive at the same results. Consequently, it is important to establish the set of correct and incorrect results in a consistent manner. Consistent means that the related results of collaborating partitions are based on the same input. Two example show how inconsistent input can lead to wrong results and hidden dependencies.

Example 1: Suppose a function is realized with three replicas that exchange their results and vote on the outcome. When all three replicas are correct, they should arrive at the same result and when a replica receives two differing results it can conclude that the result that deviates from its own result is sent by a failing replica. However, such a situation can also occur when inconsistent results are received by correct replicas. When inconsistent results are received, either an incorrect detection of a failing replica is made, or the detection of failing replicas is never done.

Example 2: Suppose a function is realized with two versions that use some similar and some different resources. Each one verifies the correctness of the input, calculates results and compares with the results of the other. Suppose the input of the common resource is different. The versions may produce inconsistent results that lead to the suppression of the whole function or may start wrong actions if one of the other unique input results is un-detectably wrong.

A requirement on the communication between resources and partitions and between partitions can be formulated:

Requirement 1. *All replicas or versions collaborating on some related results should read the same result coming from a particular resource or partition.*

Below, the failure hypotheses on the IMA hardware components are enumerated. A software architecture is proposed that satisfies the above requirement 1 and does not introduce other failure dependencies apart from the ones cited below.

3.1   Failure hypotheses

During operation of the aircraft, hardware may fail. A core module has a fail-stop behaviour: its behaviour satisfies its specification until a given moment after which the core module does not perform any observable action. When a core module fails, all the partitions executing on the core module are assumed to fail. When all processors connected to a resource fail, the connected resource is assumed to fail. When the local clock of a core module fails, the core module fails. When the power to a cabinet, all gateways or the backplane bus of a cabinet fail, the cabinet is assumed to fail. When a cabinet fails, all resources connected to this cabinet and all core modules of this cabinet fail. When the airplane bus fails, the complete avionic system is assumed to fail. A partition has fail-stop behaviour. When a partition fails, no other components of the IMA system are assumed to be affected. When the private memory of a partition fails, the partition fails. Shared memory can be divided in components with independent fail characteristics. When a part of a shared memory component fails, the whole shared memory component is assumed to fail. When all shared memory components of a core module fail, the core module is assumed to fail.

The hardware failure probabilities must be calculated such that they are inferior to the reliability requirements of the functions they support.

3.2   Communication structure

Two communication models are generally applied: (1) partition (consumer) interrogates a result producing partition or resource (producer) to return a result or (2) the producer sends at well established moment results to the known consumers. The above requirement 1 prohibits most communication models. Under model 1 several, possibly differing, results are produced by the producer. Especially when the producer can fail, some consumers receive a result and others none. For a large number of avionic functions, producers know their consumers. It is therefore preferable that producers send the same result to all consumers. When a producer sends a result to each individual consumer, the failing of the producer can lead to the unwanted situation that some consumers have received the result and others not. The best known solution is to use a reliable multicast protocol in which the producer sends one result once and the protocol guarantees that either all correct consumers receive the result or none does [5, 7]. Results are sent to processors where they are stored in shared memory at the disposal of the interested partitions. Examples of real-time system where this basic asynchronous approach is followed are: [8, 4, 10, 9]

This design satisfies the requirement 1, all producers send one result to all processors; related correct partitions will read the same results as long as their reading moment is correctly synchronized with the acceptance moment. By calculating the maximum transmission time and waiting an appropriate time with respect to the sending time, such a synchronization is achieved. When one partition of a processor fails, the results remain available in the shared memory and other partitions (in the same processor) are not affected by such a failure. When the communication fails, all involved partitions will fail as mentioned above, and when the shared memory fails, no results are available and the processor with all its partitions fails.

When producers fail not by stopping but by producing incorrect results, their results should no be accepted. Failures can be detectable on several levels.

1. The resource failure can be detectable at the resource itself by introducing redundancy checks at the source.

2. A software layer on top of the physical resource can check the correctness of the produced results.

3. A test program can asynchronously with the control programs access resources and check their correct functioning.

4. A partition can conclude that a result is incorrect by comparing the result with results from functionally equivalent producers.

5. Core modules can detect that a core module or transmission channel fails.

In cases 1 and 2, the correctness of a result can be determined before the result is multicast to its consumers. By adding the status of the resource to the produced result, all partitions take the same decision on the refusal or acceptance of the requested results. However in cases 3-5, a distribution of failure information that is synchronous with the result distribution is much harder to realize. An example will show this.

Example 3: Suppose a result is sent at time t_s and received and stored on all processor memories at time t_r. A partition p will read these results at time $t_p > t_r$. Suppose a failure is detected in the result producing resource and is communicated to all processors at time $t_f > t_r$. It is easy to see that the partitions cannot read the result at the same time t_p and that for some partitions $p : t_p < t_f$ and for other partitions $p : t_p > t_f$, thus violating the consistency requirement unless special measures are taken.

The membership service can help to solve this problem. The realization of the membership service [6] extended to hierarchical communication structures [11] and adapted to avionics [12] is described below.

4   MEMBERSHIP

A consistent distribution of failure information is formulated in the following requirement:

Requirement 2. *All replicas or versions collaborating on some related results have the same opinion on the correctness of a particular resource or partition.*

By considering the set of correct resources and partitions, the above requirement can be reformulated as the the membership requirement by asking that every partition has the same view on the members of the set of correct partitions and resources. The normal design approach is to assure that a processor has a view on the membership set and all correct partitions executing on this processor share the same view. The moments that a view of a given processor changes is defined with respect to the local time of the processor. A membership algorithm realizes this common view on the membership set. The requirements on the membership service are:

Requirement 3. Membership

- *At identical local clock times, the membership view of any two correct processors is identical.*

- *Resources and partitions that are detected to fail by a processor p at local time t are removed from the membership, set within a bounded period J, at local times $t + J$.*

- *Correct resources and partitions are not removed from the membership set.*

The storage of a new result in shared memory depends on the presence of the producer in the membership set. Assume a constant K such that a sent result arrives at the destination processor p at a local time $t_p < t_s + K$. When a result -sent at local time t_s- arrives in a processor at local time t, the result is stored in the shared memory of the receiving processor at local time $t_r = t_s + K$ if the producer of the result is a member of the membership set at local time t_r. The realization of the membership algorithm and the calculation of constants K and J is the subject of the next subsection.

4.1 Membership realization

It is assumed that during any moment any two correct processors can communicate with each other. This assumption is justified by the failure hypothesis that the failure of the communication network implies the failure of the complete IMA system. It is assumed that the failure detection of resources does not detect failures for correct resources. Problem is to determine that no correct cabinets or processors with their connected resources are removed from the membership set. When the mebership set is maintained in a reduced set of processors, a larger communication overhead is needed and the membership service can be lost with still functioning processors available. As long as there are correct processors, the membership algorithm needs to satisfy requirement 3. Therefore, centralized solutions are excluded. Replication of the membership state over all correct core modules is a consequence. When a partition, core module or cabinet fails, the membership state remains updated in all surviving core modules.

The membership algorithm consists of a *detection* part and a *distribution* part. The failures of processors and partitions are detected by having each processor send message to ascertain its correctness. The messages contain information about failed resources and partitions to distribute the failure information. Partitions that detect the failures of other partitions or resources communicate this to the membership service that is responsible for the timely distribution. The moment that failure information is communicated to the membership service, is defined as the failure detection time.

Design Partitions are responsible for monitoring the resources. Every cabinet is responsible for monitoring the state of the partitions and core modules. The membership algorithm is periodically executed. Four phases are discerned: a gathering phase, a collection phase, a distribution phase and an information phase.

In the *Gathering* phase, partitions actively inform the membership service of resource failures and partition crashes by invoking functions that store this information in the shared memory (not the membership state) of the core module hosting the detecting partition.

In the *Collection* phase, the core modules communicate the failure information at a prearranged local time to all other core modules in the same cabinet via the cabinet bus. When no failure information is present, the core module sends an empty message. The failure of a core module to send a message is interpreted by the other core modules as a core module failure. To assure that all correct core modules within the cabinet receive the same message or none, a reliable multicast is needed as already mentioned above for the transmission of results by producers. When a core module does not send, the core module has crashed, its shared memory has crashed, or its local clock has crashed. This is in accordance with the hardware failure hypothesis.

In the *Distribution* phase, all cabinets distribute the cabinet's failure information to all other cabinets via the airplane bus. When no failure information is present, the cabinet sends an empty message. The meader/shadow mechanism is used to communicate the failure information. All correct core modules in a given cabinet have the same failure information. Consequently, the master and shadows will send the same information to other cabinets. All core modules of the destination cabinet receive the message. When a cabinet does not send, either all its gateway modules, its power-supply, its backplane bus or the master and all shadows have failed. This is in accordance with the hardware failure hypothesis.

In the *Information* phase, all correct core modules install the new membership state at the same local clock times. All correct core modules have received the same failure information. All correct core modules had the same membership state. Consequently, after the information phase all core modules have the same membership state. When the membership state changes, all correct modules change state at the same local clock time.

Performance To satisfy the real-time criteria, a well defined time bound between message transmission and message reception must be assured. The Airplane bus gives such a bound only for periodic messages. Consequently, every cabinet transmits its view periodically every π time units.

From the communication descriptions it can be concluded that the communication within a cabinet is strictly synchronized, but communication between cabinets is sensitive to clock drifts. The timing of the message transmission between cabinets drifts with respect to the cabinet backplane bus schedules. This adversely affects the transmission delay.

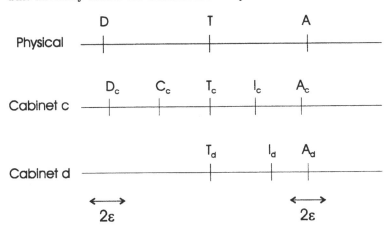

Figure 4 Communication delay for membership

A timing diagram is shown in Fig. 4. The period of the airplane data bus π is smaller than the period π_c of the backplane bus of any given cabinet c. Every meader/shadow transmits the cabinet information $S_c = \lceil \pi_c / \pi \rceil$ times. Suppose a core module in cabinet c detects a failure at local time D_c. During the collection phase, this core module transmits the failure on the cabinet data bus at local time C_c. The collection phase starts after the detection phase with a maximum delay given by: $D_c + \pi > C_c > D_c$. At local time T_c, determined by the timing of the airplane bus, the results are transmitted to the receiving cabinets. Neglecting the transmission delay, the message arrives at cabinets c and d at global time T, that is the same for both cabinets as they are connected by the same airplane bus. Although, transmission occurs at the same physical time, the local clocks T_c and T_d differ by a value ϵ: $\mid T_c{-}T_d \mid < \epsilon$. The periodicity of the airplane bus introduces a maximum delay of π: $T_c - \pi < C_c < T_c$. A message sent at local time T_c is received at times L_c and L_d by core modules in cabinets c and d with the restraint $(x = c \text{ or } x = d) : T_x < L_x < T_x + \pi$. When both cabinets follow the same frame organization, the values of L_c and L_d are identical but have a spacing of approximately ϵ time units. The acceptance of the message by a core module in cabinet d is done at time A_d such that $\forall x : A_d \geq L_x$. This leads to: $T_d + \pi < A_d < T_d + 2 \cdot \pi$. According to the membership requirements, the values A_c and A_d are identical but occur at different physical times placed within an interval of approximately ϵ. Expressed in local times the difference between D_c and A_d is given by:

$$\pi - \epsilon < A_d - D_c < 3 \cdot \pi + \epsilon \qquad (11.1)$$

Assuming that clocks drift little with respect to the physical clock in interval ϵ, The following constraints are true:

$$|D - D_c| < \epsilon \quad |A - A_c| < \epsilon$$

The total physical time is given by:

$$\pi - \epsilon < A - D < 3(\pi + \epsilon) \qquad (11.2)$$

5   CONCLUSIONS

Avionic functions are realized with an IMA architecture. The strict hardware separation of functions of different criticality levels is no longer an interesting option. Resources are shared and especially the communication media between different core modules must be shared. Important is the prevention of failure propagation between components with a low criticality level to a high criticality level. It is argued that partitions (versions or replicas) that collaborate to produce the same results need the same input data. A communication structure is defined that satisfies this requirement and depends on the availability of a reliable multicast facility with bounded transmission times. It is shown that the hardware determined failure hypotheses are not extended by this structure.

The introduction of failures and the requirement that collaborating partitions consider the same resources as failed at the same moment can be realized with a membership algorithm. An outline of the algorithm is presented and it is shown how the failure hypotheses are maintained. Results are only possible if bounded transmission times are possible. The required bounds are presented as function of the IMA communication media characteristics.

Acknowledgements

We are grateful to Paul Weir from British Aerospace and René Meunier from Aérospatiale for many stimulating and helpful discussions.

6 REFERENCES

[1] Committee Arinc 629. ARINC report 629, Multi-Transmitter Data Bus. Technical Report Specification 629, Airlines Electronic Engineering Committee, March 1990.

[2] Committee Arinc 651. ARINC report 651, draft 9. Technical Report 91-207/SAI-435, Airlines Electronic Engineering Committee, September 1991.

[3] Committee Arinc 659. ARINC report 659, draft 4. Technical Report 92-259/SAI-477, Airlines Electronic Engineering Committee, October 1992.

[4] M. Boasson. Control systems software. *IEEE Transactions on Automatic Control*, 38(7):1094–1106, 1993.

[5] F. Cristian. Atomic Broadcast: From simple message Diffusion to Byzantine Agreement. In *Proceedings 15th International Symposium on Fault-Tolerant Computing*, pages 200–206, Ann Arbor, MI, June 1985.

[6] F. Cristian. Agreeing on who is present and who is absent in a synchronous distributed system. In *Proceedings 18th International Symposium on Fault-Tolerant Computing*, pages 206–211, Tokyo, Japan, June 1988.

[7] F. Cristian. Synchronous Atomic Broadcast for Redundant Broadcast Channels. *Journal of Real-Time Systems*, 2:195–212, 1990.

[8] T Kuo and A.K. Mok. Application Semantics and Concurrency Control of Real-Time Data-Intensive Applications. In *Proceedings of 13th Real-Time Systems Symposium*, pages 35–45, Phoenix, December 1992. IEEE.

[9] R. Meunier. Airbus architectures. private communication.

[10] K. Ramamritham. Real-Time Databases. *Distributed and Parallel Databases*, 1993(1):199–226, 1993.

[11] P.D.V. van der Stok, M.M.M.P.J. Claessen and D. Alstein. A hierarchical membership protocol for synchronous distributed systems. In *Proc. of the*

1st Eropean Dependable Computing Conference, pages 597–616. Springer Verlag, October 1994.

[12] P. Weir and P.D.V. van der Stok. Task 2: Requirements on Cabinet monitoring aspects in IMA context. Technical Report SDF/B67/A/108/2308, Brite-Euram, May 1994.

12

PETRI NET MODELING AND BEHAVIORAL FAULT MODELING SCHEME FOR VHDL DESCRIPTIONS

D. Fédérici, J-F. Santucci, P. Bisgambiglia

Fédérici Dominique, Santucci Jean-François, Bisgambiglia Paul
Université de Corse Quartier Grossetti BP 52, 20250 Corte, France
e-mail: federici,santucci,bisgambi@univ-corse.fr

ABSTRACT

This paper deals with the modeling of VHDL behavioral descriptions and the development of an efficient behavioral fault modeling scheme. A set of behavioral test pattern generation methods have been proposed in the recent past. One constraint having to be pointing out is the fact that circuits under test (CUT) are expressed using a high level behavioral VHDL description. We propose to define a behavioral fault simulation method own to (i) a behavioral modeling of CUT using Petri Nets and (ii) an efficient behavioral fault modeling scheme. In this paper, the emphasis is put on the modeling aspects.

1   INTRODUCTION

In the recent past a set of Behavioral Test Pattern Generation methods have been proposed in several international conferences [1, 2, 3, 4, 5, 6, 7, 8]. Motivated by this fact we are interested in defining an efficient behavioral fault simulation methodology. One important constraint within which we have to work is the fact that circuits under test (C.U.T.) are described using a high-level behavioral description. The previous work has focused on the A.T.P.G. method and not on the fault modeling and simulation tasks.

Our goal in this article is to present an efficient modeling and fault modeling scheme for behavioral descriptions.

The first part of the presentation is devoted to general definitions of modeling and fault modeling. The second part will deal with behavioral modeling of VHDL descriptions. The definition of an exhaustive behavioral fault modeling scheme is given in the third part. Future work concerning behavioral fault simulation is briefly introduced in the last section.

2   GENERAL DEFINITIONS

The modeling of a circuit can be done according to two points of view: a structural view and a behavioral one. We focus on discrete models. The use of a discrete model to solve a given problem refers to a set of discrete variables called State Variables which define the State Space of the model. The behavior associated with the model is necessarily expressed through the alteration of the variable values.

In case of a behavioral view, the circuit is seen as a black box defining its output values according to input values by the use of algorithms, true tables, state tables or boolean equations.

Behavioral models can have two main representations:

- alphanumeric representations: they are textual representations involving objects such as variables, operators and control constructs.

- graph representations: they are based on transformation graphs, Petri Nets, etc... . These representations offer a structure, i.e. an interconnexion of basic elements for which a behavior is predefined out of context. We have to point out that this structure is not a potential structure but some elements can have a physical interpretation.

Whatever the general principle on which the Test Pattern Generation (T.P.G.) or Fault Simulation (F.S.) methods are based, structural or behavioral models of the circuit under test are considered as the reference or faultless model.

Given a description (structural or behavioral view) of a circuit, the T.P.G. or F.S. method has to operate on the information available by the description. In both cases fault models are needed in order to represent physical failures efficiently.

Before dealing in detail with behavioral fault modeling, we propose some definitions of a fault, a fault model, an error and a defect in order to guide the reader towards behavioral fault modeling.

A fault can be defined both on a structural and a behavioral model. In the case of the use of a discrete event model, a fault hypothesis is:

- either an hypothesis of a wrong item behavior belonging to the model, but considered out of context.

- or an hypothesis of modification of the initial global description by adding, suppressing or combining basic items, without modifying the predefined behavior of these items.

A fault model is the list of the selected fault hypothesis, i.e. the fault hypothesis taken into account for T.P.G. or F.S. There are several interests in using a fault model. The main one is to reduce all possible combinations of T.P. sequences at the input of the circuit under test. The definition of fault hypothesis allows the definition of fault classes, the reduction of the list of selected faults. Furthermore, a fault simulator can be used to determine all the faults set off by a given test pattern. Lastly, an hypothesis may be used to make a correlation between a fault and a physical defect which is useful for fault localization.

It should be pointed out that the modification of the global description of the model may lead to take into account a much too large number of possible combinations. In a similar way, the overly complex faults concerning the modification of a basic items behavior may also not be taken into account.

Whatever the given abstraction level, an error is the manifestation of a fault expressed as a difference between the state of the fault free model and the faulty model at a given time. If it is not posssible to obtain a difference between the two models, the fault is said to be redundant.

A defect refers to a physical anomaly of the actual circuit. It should be pointed out that a fault may or may not have a direct mapping with a physical defect.

3   BEHAVIORAL MODELING

In order to facilitate definition of formal methods based on graph structures, we have been interested in a first step to study an efficient graph modeling of behavioral descriptions. Behavioral descriptions are given using the VHDL language. In a first sub-section we highlight the main features of VHDL behavioral descriptions. We describe in a second sub-section how these features have been represented by an efficient graph modeling scheme.

3.1   Main features of VHDL Behavioral Descriptions

Four kinds of objects are involved in VHDL behavioral descriptions:

- Constants which have a predefined and unchangeable value.

- Variables which can be modified by an assignment statement.

- Signals which are the specificity of VHDL.

- Processes which are the fundamental objects manipulated by VHDL.

Variables are local to processes that is to say they can be read or affected only in the process where they are declared.

Signals are global to the overall description that is to say that they may be common to several processes.

A behavioral description leans on the description of a set of processes where each process is defined using a procedural description.

The key statement of a process is the WAIT statement.
syntax: WAIT {ON signal-list} {UNTIL boolean-condition}

This statement allows to suspend a process execution, that will restart when the next condition will be true: an event occurs on one of the signals specified in the signal_list and the evaluation result of the boolean_condition is true.

The simulation of a VHDL description is composed of two steps: an initialization phase and an execution phase.

The initialization phase consists in determining the initial value of each signal according to the following rules:

- the initial value is given explicitly during the signal declaration.

- the initial value is given implicitly. This value is defined like being the first of the signal definition domain.

With these initial values, each process of the description is executed without looking at the conditions associated with Wait statements.

The execution phase is performed by scanning of the sensitive signals and is driven by the events.

This modeling scheme will be described in sub-section 3.2 using the behavioral description given in Figure 1.

3.2   Behavioral Modeling

This sub-section aims at describing the model used to highlight the main features of VHDL behavioral descriptions. In order to facilitate definition of formal methods based on graph structures for behavioral fault simulation, we have been interested in the development of two models:

- the input/output model allowing to represent existing links between signals involved in different processes of a description [9]. This model has been defined in [9] in order to perform Test Pattern Generation.

- the activation model pointing out the the execution of the processes involved in a VHDL description. This model is based on a model developed in [9]. However we have improved this previous model by defining a new modeling scheme involving pure Petri Nets. The use of Petri Nets will be useful when a behavioral fault simulation algorithm will be defined.

The VHDL behavioral description activation model is given in Figure 2. We have to mention that the activation model is made up of the following elements: a data model, a control model and their explicit interactions. The Data Model

```
Entity Register IS
Port (DI   IN vlbit_1d(1 TO 8) ;
          STRB, DS1, NDS2 : IN vlbit ,
          DO   OUT vlbit_1d(1 TO 8)) ,
END Register

Architecture behavior of Register IS
SIGNAL reg  vlbit_1d(1 TO 8);
SIGNAL enbld   vlbit ;
BEGIN

    strobe: PROCESS (STRB)
    begin
    If (STRB = 1) then reg <= DI;
    End If,
    END PROCESS Strobe,

    enable : PROCESS (DS1,NDS2)
    beign
    enlbd <= DS1 AND NOT (NDS2);
    END PROCESS;

    output : PROCESS (reg,enbld)
    Begin
    If (enbld=1) then DO<=reg
    Else DO  <= 11111111,
    end If
    END PROCESS,

END Behavior
```

Figure 1   VHDL behavioral description

represents the objects (variables and signals) and the handled operations involved in the description. It is composed of a graph involving two types of
nodes: data nodes and operation nodes. Let us point out that the data nodes
have not been represented on Figure 2. Operation nodes are of two types:

- assignment nodes which represent the assignment of an object by an algebraic or boolean operation.

- decision nodes which represent an algebraic or boolean test the result of
  which is taken into account in a branch in the Control Model.

The Control Model represents the sequencing of the operations involved in the
description. It is based on a Petri Net modeling. The interaction between the

Control Model and the Data Model is achieved by associating an operation node to a place. Two links are involved between the place and the operation node: an activation link of the operation node and an end report link. The dynamical aspect of the interaction of the two models is supported by tokens. When a token arrives in a place, the associated operation is performed. In case of a decision node, the result associated with the end report link is used in order to select the next transition to fire. In Figure 2, we give the Petri Net associated with the description given in Figure 1.

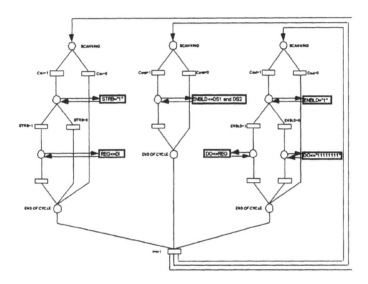

Figure 2 Activation Model

4 BEHAVIORAL FAULT MODELING

A fault modeling scheme based on the definition presented in section 2 allows to derive a set of fault hypotheses on the previously described model. Fault hypotheses are defined according to the elements involved in the model of the circuit under test. A General Fault Model (G.F.M.) can be generated using the two rules presented in section 1. This fault Model involves all possible fault hypotheses defined on the basic modeling element and with modification of the model structure. Because of the too large number of faults involved in this model we may have to reduce it. In order to have a behavioral fault model for which some measures of confidence are provided, we may select from among the G.F.M. some fault hypotheses according to the fault model proposed in [10]. The selected fault hypotheses may be classified as follows:

1. Stuck-at fault on an element of the data model:

   – The value attribute of a data node is stuck-at V1 or V2 where V1 and V2 express the lower and upper extremes of the domain definition of the represented signal or variable.

   – The output of an operation node may fail such that it permanently returns V1 or V2, where V1 and V2 express the lower and upper extremes of the range of the operation.

2. Stuck-at fault of an element of the control model:

   – a control transition is always selected whatever the resulting value set up on the end report link may be

   – a process is never active

   – a process is always active

3. Stuck-at fault of an element of the interaction of the control and data models:

   – when a token arrives in a place, the corresponding operation is not performed. This means that the activation link is stuck-at 0.

Depending on the results given by these fault hypotheses in terms of gate-level fault coverage, this fault model can obviously be extended. However, having translated the behavioral fault model proposed in [10] by a set of fault hypotheses on the basic elements of the activation model defined in section 2, the quality of the test vectors generated for the previous fault model can be evaluated by the results presented in [10].

5   CURRENT AND FUTURE WORK

Our current work deals with behavioral fault simulation [11]. Fault simulation is known to be an important step in the testing of digital systems. It is usually used to evaluate the fault coverage of each test vector generated by an Automatic Test Pattern Generation program. Traditional fault simulation algorithm [12, 13, 14] are not able to deal with behavioral description and behavioral fault hypotheses. We are interested in defining new algorithms allowing to perform fault simulation on VHDL behavioral descriptions. Given a test sequence, a VHDL description, a set of behavioral fault hypotheses, our goal is to deduce the list of behavioral fault hypothesis detected by the given test sequence. The first approach we have investigated is to define an algorithm based on the deductive fault simulation [12]. This approach consists in propagating a fault list through the basic elements of the activation model. Given a test sequence, this propagation is made until that we meet a primary output. To evolve this method, we have decomposed our activation model as an interconnexion of four basic elements shown in figure 3.

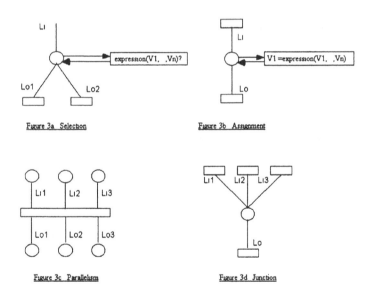

Figure 3 Basic elements involved in the activation model

We have defined how lists of detected faults Loi can be computed according to:

■ each kind of basic element,

■ Lii, list of detected faults,

- a given test pattern.

Our current work is to find out, how a fault list is propagated through each basic element. Our future work will investigate the definition of a method based on a concurrent fault simulation [14].

6   REFERENCES

[1] D.S. Barcaly, J.R. Armstrong, "A chip-level Test Generation Algorithm", 23th IEEE/ACM Design Automation Conf. 1986, pp.257-261.

[2] U.H. Levendel, P.R. Menon, "Test Generation algorithms for computer hardware description languages", IEEE Transactions on Computers, Vol.C-31, pp.577-588, 1982.

[3] M.D. Oneill, D.D. Jani, C.H. Cho, J.R. Armstrong, "BTG: a Behavioral Test Generator", 9th Computer Hardware Description Languages and thier Application, IFIP, pp.347-361, 1990.

[4] F.E. Norrod, "An automatic test generation algorithm for hardware description language", 26th Design Automation Conference, pp.76-85, July 1989.

[5] H.D. Hummer, H. Veit, H. Toepfer, "Functional Tests for hardware derived from VHDL description", CHDL 91, pp. 433-445, 1991.

[6] A.L. Courbis, J.F. Santucci, N. Giambiasi, "Automatic Test Pattern Generation for Digital Circuits", 1st Asian Symposium, Hiroshima, pp. 112-118, 1992.

[7] J.F. Santucci, A.L. Courbis, N. Giambiasi, "Behavioral Testing of digital Circuits", Journal of Microelectronic Systems Integration, Vol.1, N 1,, March 1993, pp.55-78.

[8] B. Straube, M. Gulbins, E. Fordran, "An approach to Hierarchical test Generation at Algorithmic Level", IEEE Workshop on hierarchical test Generation , Blackburg, Virginia, USA, 8-11 august 93.

[9] V. Pla, J.F. Santucci, N. Giambiasi, "On the modeling and Testing of VHDL Behavioral Descriptions of Sequential Circuits, 3rd IEEE Euro-Dac/Euro- VHDL, Hamburg, September 93, pp.440-445.

[10] S. Ghosh, T.J. Chakraborty, "On behavior fault modeling for digital designs", Journal of Electronic Testing: Theory and Applications, Vol. 2, pp. 135-151, 1991.

[11] D. Fédérici, J.F. Santucci, P. Bisgambiglia, "Behavioral Fault Modeling and Simulation", 3rd BELSIGN Workshop Proceedings, 11-12 April 96, Porticcio, France.

[12] D.B. Armstrong, "A Deductive Method for Simulating Fault in Logic Circuits", IEEE Trans. on Computers, Vol C-21, N 5, May 1972, pp. 464-471.

[13] P. Goel, P.R. Moorby, "fault Simulation Techniques for VLSI Circuits", VLSI Design, July 1984, pp.22-26.

[14] E.G. Ulrich, T.Baker, "The Concurrent Simulation of Nearly Identical Digital Networks", Proc. 10th Design Automation Workshop, IEEE and ACM, New York, June 1973, pp.145-150.

This work is supported by the EEC: HCM network BELSIGN- contract N CHRX-CT 94-0459.

CATSAT'S SOFT X-RAY DETECTION SYSTEM: AN INNOVATIVE AND COST EFFECTIVE APPROACH

S. P. Lynch

Small Satellite Laboratory
Institute for the Study of Earth, Oceans & Space
University of New Hampshire
Durham, NH 03824, USA

ABSTRACT

This paper describes an innovative and cost effective approach to detecting low energy x-ray emissions from gamma-ray bursts. This data will assist scientists in solving the important problem of locating the source of these bursts. The approach makes the low energy measurements using an array of avalanche photodiodes (APDs). APDs are a proven, inexpensive technology that will enable the system to be flown aboard a small satellite, CATSAT.

1 INTRODUCTION

The Cooperative Astrophysics and Technology Satellite (CATSAT) is a combined effort between three universities to study the origin of cosmic gamma ray bursts. Astrophysicists have speculated for years as to how far away the bursts occur, but to date no common theory is agreed upon. In a unique approach to solving this problem CATSAT will measure low energy x-rays, known as "Soft X-Rays", emitted from the bursts using an array of APDs, and use this data to calculate how far away the bursts occurred. The APDs' relative noncomplexity and low cost enables such an important scientific contribution to be made using a small satellite mission.

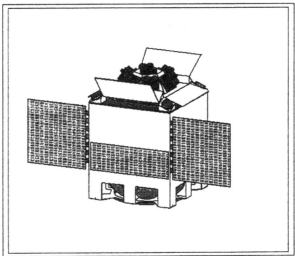

Figure 1: CATSAT Small Satellite

Supported by a grant from NASA's Student Explorer Demonstration Initiative, (STEDI) CATSAT is representative of the emerging field of "small satellites" [6]. Small satellites are typically less than 600 kg in mass, fly in a low earth orbit (550 km) and have a design time of two years. Economic constraint is the driving force behind the development of small satellites, as larger missions become increasingly complex and expensive. Student involvement in the design and manufacturing processes is an integral part of the STEDI program and students from all three universities are playing an active role as CATSAT's design engineers. CATSAT is currently scheduled to be launched in July of 1998.

This paper focuses on the Soft X-Ray (SXR) detection system for CATSAT, which includes the APDs and their associated analog and digital circuitry. The SXR system is contrasted with conventional low energy x-ray detection systems, which would not be feasible for a small satellite mission due to their cost and complexity.

2   CATSAT ORGANIZATION

The CATSAT team is comprised of three universities, the University of New Hampshire (UNH), Weber State University (WSU) in Utah and Leicester University (LU) in England. At UNH, the mission leader, the Institute for the Study of Earth, Oceans and Space (EOS) is collaborating with the Electrical and Computer Engineering Department (ECE). This team will design the spacecraft's scientific instrumentation and electronics. Much of the design work is being carried out by undergraduate and graduate students as per STEDI guidelines. The instrument design group at UNH is working closely with Leicester University in England, where the housing for the SXR detectors will be built and tested.

The Center for Aerospace Technology (CAST) at Weber State University in Utah, is responsible for the design and fabrication of the spacecraft frame. WSU is also responsible for spacecraft attitude determination and control.

3   SOFT X-RAY DETECTION SYSTEM

3.1   Scientific Basis

Gamma-ray bursts (GRBs) were first detected in the early seventies. Vela satellites accidentally discovered them as they were monitoring Soviet compliance with the Nuclear Test Ban treaty [2]. Since that time astrophysicists have been unable to satisfactorily explain both the nature and the origin of GRBs. Current theories vary widely, from local origins to galactic and even extragalactic origins. It is generally accepted that the bursts are caused by extremely energetic events which may be some of the most spectacular in nature. Studies have been unsuccessful in trying to correlate the 400 bursts reported per year with events at different wavelengths or other astronomical objects. It was hoped that counterparts to the GRBs would provide clues as to their origin using searches at radio, infra-red, optical and x-ray wavelengths [5].

CATSAT's innovative, multi-observational approach will give scientists the first measurements of the distance to gamma-ray bursts, solving the most important problem in GRB studies. The key to these distance measurements lies in the SXR system's ability to measure low energy x-ray data.

3.2   SXR Module

The SXR module is located on the top of the satellite where it obtains a field of view of approximately 2pi steradians. It is divided into 7 independent panels, each one containing 16 APDs, providing a total of 112 APDs. The module housing is designed to radiate heat and maintain the temperature of the APDs at 40 C, ± 15 degrees. Protective doors cover the module during launch and are opened once the satellite has maintained a stable orbit.

The instrument electronics for each of the 7 panels amplifies the APD signals, digitizes the data and temporarily stores it in memory. The data is systematically retrieved from the panel memory by the instrument CPU, stored in the main memory and downloaded to earth every twelve hours. The in-flight calibration (IFC) circuitry for each panel will maintain the APDs at stable operating conditions.

3.3   Avalanche Photodiodes (APDs)

The APD is ideal for low energy x-ray detection in a small spacecraft. It is a compact, low power device and has excellent energy resolution. In-flight calibration circuitry will individually gain adjust each APD. This stabilizes the operating characteristics of all the APDs in a panel for example, and ensures that the spectra collected with that panel will appear as though it was taken with one large device.

The APD used aboard CATSAT is a silicon device with a thickness of about 2 mm and a detector top surface area of 1.69 cm2. The APD cutaway view, shown in Figure 2, displays the p diffusion region on top, the multiplication region in the middle and the n diffusion region on the bottom. The diagram depicts the operation of the APD, whereby a charged particle strikes the p-region causing an avalanche process that produces a current pulse at the output.

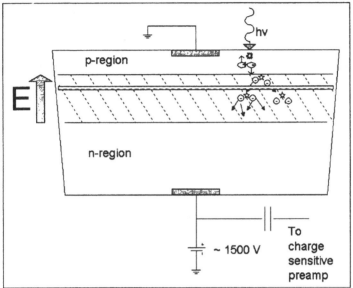

Figure 2: Avalanche Photodiode

4 ECONOMIC CONSTRAINTS

STEDI funding provides a total of 4 million dollars to the CATSAT project for design and fabrication of the entire spacecraft and its payload. This is modest in comparison to the cost of many modern spacecraft, which can be 100 times more expensive and even then are not guaranteed to be successful. The cost of launching "traditional" satellites can also be enormous. NASA and other governmental agencies across the world realize this and recent years have seen a concerted effort to promote small satellite design as a valuable alternative to large missions. In fact, it is a specific goal of the STEDI program, which was initiated by NASA and the Universities Space Research Association, to "demonstrate that small, relatively low-cost, and rapidly developed space missions can both enrich education and produce significant science" [6].

It is with respect to these economic constraints that a tremendous advantage of the SXR system can be seen; its total cost is a fraction of that of conventional methods of low energy x-ray detection. Without this new application of APDs in the SXR system it is doubtful that CATSAT would be able to complete its mission given the limited budget. Because of the relatively low cost of the APDs (approximately $1,000 each) the complete array for the SXR system can be built for just over $100K. The electronics, which includes low noise "hybrid" devices, is estimated to cost between $50K and $60K, and the housing can be built for under $10K. This allows the complete system to be built for around $200K.

Table 1 below shows a comparison of the estimated cost of the SXR system compared to other methods that could be used. Note that the estimates are based on a system that would have a comparable field of view to the SXR system.

5 SYSTEM COMPLEXITY

System complexity is an important factor in a small satellite mission. The two year design and construction period along with the limited payload capabilities of a small spacecraft would make it difficult to use the conventional methods mentioned in Table 1.

System	Cost ($)
Soft X-Ray System	200K
Low Energy Photon Spectrometers [4]	5M
Charge Coupled Device	1M

Table 1: Cost Comparisons

Low energy photon spectrometers for example, operate at extremely low temperatures, requiring the storage of liquid nitrogen onboard the spacecraft. The surface area of the detectors is also very small, approximately 10 mm square, which would require the use of a large number of them to obtain a sufficient field of view. Charge Coupled Devices (CCDs) are also impractical as they have a "dead time" that would severely hinder the ability of the instrument to record bursts.

While the SXR system does require precise control of the APDs, it is still less complicated in comparison to the above methods. The system incorporates passive environmental control and also has built in redundancy due to the large number of detectors.

6 CONCLUSION

This paper has described an innovative approach to solving a problem that has long been debated by the astrophysics community: the origin of gamma-ray bursts. Furthermore, the SXR detection system demonstrates that spacecraft technology can be developed at low cost and in a short time frame. The CATSAT mission has found a new use for an existing technology, and incorporated it in a small satellite program for a fraction of the cost of conventional systems.

The relative noncomplexity of the SXR system has allowed undergraduate and graduate engineering students to work on a real-life design effort and gain valuable experience. If successful, CATSAT will be a powerful incentive for "smaller, faster and cheaper" missions that can provide important contributions to the scientific community.

7 ACKNOWLEDGEMENTS

The author would like to thank the CATSAT project for graduate funding. In addition, Dr. Ron Clark, Dr. Andrzej Rucinski, Dr. David Forrest, Dr. Tom Vestrand, Ken Levenson and Glenn Forrest are thanked for their help and mentoring throughout the course of this work.

8 REFERENCES

[1] Farrell, R., Vanderpuye, K., Entine, G. and Squillante, M.R. High Resolution, Low Energy Avalanche Photodiode X-Ray Detectors. IEEE Transactions on Nuclear Science, Vol. 38, No. 2, April 1991. 144-147.

[2] Klebasadel, R.W., Strong, I.B. & Olsen, R.A., Observations of Gamma-Ray Bursts of Cosmic Origin. "The Astrophysical Journal," 1973, 182, L85.

[3] Knoll, Glenn F. Radiation Detection and Measurement, John Wiley & Sons, New York, 1989.

[4] EG&G Ortec, Modular Pulse-Processing Electronics & Semiconductor Radiation Detectors, Product Catalog, USA, Nov. 1994.

[5] Owens, A., Schaefer, B. and Sembay, S. Deriving Gamma-Ray Burster Distances from Soft X-Ray Measurements. "The Astrophysical Journal", In-Press, 1995.

[6] "STEDI - World Class Orbital Science by Students", USRA QUARTERLY. Special Edition, December, 1994.

14

PETRI NETS FOR A SPACE OPERATIONAL SYSTEM AVAILABILITY STUDY

M. Saleman*, J-F. Ereau**

*CNES, 18, avenue Edouard Belin 31055 - Toulouse
e-mail: malecka.saleman@cnes.fr
**CNES & LAAS-CNRS Toulouse

ABSTRACT

This paper presents an availability study of a space operational system called TPFO (Topex Poseidon Follow On) using Petri nets for both modeling and evaluation aspects. The aim of this study was to examine mission/system trades that might be possible in view of TPFO missions with the intent of achieving minimum program cost. Thus, the study have to take into account several aspects of the program, as mission time, number of satellites, satellite reliability and lifetime, satellite production and storage policies, launch reliability and availability, relaunch policy, etc... and to identify options for achieving availability objectives with minimum total cost. We illustrate in this paper Petri net's ability to deal with such system from both modeling and evaluation view points. The results obtained by this method provide interesting outputs for the project. This is shown through the interpretation of typical results.

1 INTRODUCTION

Today space projects are far from their initial experimental feature and in many application areas, such as communication, localization or observation, these systems have to deal with operational constraints and minimum program cost. Among the various system analyses performed during the early design phases of projects, availability analysis provides important results for system dimensioning. Actually, they help to select maintenance strategies and associated resources, to satisfy availability objectives with minimum or reduced global cost. Considering, the modelisation view points of such systems availability are confronted with several constraints like sequential processes as satellites

191

productions, parallel processes as satellite productions and launch requests, synchronized processes as launch campaign needed a satellite and a launcher available. In addition, different time characteristics have to be taken into account: deterministic ones like satellite production time or satellite lifetime, and stochastic ones like satellite time to failure. Petri nets are well adapted to support a structured approach because they are based on a system view under the form of a set of sequential processes or active objects which are interacting. This is the reason they offer an easy understanding of the system behavior which can improve the dialog between the analyst and the project team. In addition, the Stochastic and Time Petri Net (STPN) (Ref. 1), which is an extended model of Generalized Stochastic Petri Net (GSPN) (Ref. 2), is able to deal with arbitrary time distribution like exponential, deterministic or uniform. Hence, performance parameters, like availability, in the transient and stationary periods can be estimated by the simulation of the nets.

The aim of this paper is twofold; through an example of a satellite system, we illustrate Petri Net ability for modeling and evaluation of such systems, and thanks to interpretation of typical results, we stress how such availability studies can be useful for project management. Section 2 presents the system example to be studied. Input parameters over which sensitivity analysis can be performed are defined. Section 3 presents the Petri Net models built to represent this system. We focus here on the description of the logical behavior of the system, hence the time parameter is not explicitly taken into account. Section 4 justifies the choice of the time extension of the Petri Net used: the Stochastic and Time Petri Nets. We present the principle of the net simulation and, for a specific value of the defined input parameters, we provide and comment on typical results. Section 5 emphasizes the usefulness of such a flexible approach from both the analyst and the project manager view points.

2   SYSTEM DESCRIPTION

The system is described by its input parameters and its logistic support policy.

2.1   Input parameters

The system studied is composed by (see figure 1):

- the space segment, and

- the ground logistic support segment, i.e. all the resources needed to set up and maintain the space segment

The space segment is based on a Low Earth Orbit (LEO) satellite
The ground logistic support includes the following resources:

- a satellite production line,
- the possibility of a stock capacity
- a launching area

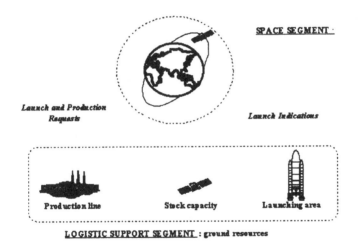

Figure 1. System components

Table 1 provides the stochastic and time input parameters for with both space and logistic support segments. The failure rates for nominal satellite is supposed to be constant, so satellite reliability is exponentially distributed. The values of these input parameters come from historical technical characteristics of space vehicles, launchers, and production systems.

Parameter	Comments
SPACE SEGMENT	
$\lambda(sat)$	Satellite failure rate
$T_{life}(sat)$	Time to satellite end of life
T_{test}	Time injection and test a satellite
$P_{injection}$	Probability of injection success
LOGISTIC SUPPORT SEGMENT	
$T_{prod}(sat)$	Time to produce a satellite
$T_{avail}(launch)$	Time to have a launcher available
$T_{campaign}$	Time of the launch campaign
P_{launch}	Probability of launch success

Table 1. Time and stochastic input parameters

2.2 Initialization and Maintenance policy

Initially, a satellite is produced and a launcher is ordered. As soon as a launcher and a satellite are available and the campaign is done, the satellite is launched. If the launch fails or the injection is not successful, a request of a satellite and 1 launcher is made and a new launch is programmed as soon as possible. If the launch and the injection are successful, the satellite is operational on orbit until his failure or end of his lifetime. Before the end of life of the operational satellite in orbit, a satellite production and a launcher request are programmed to have them available to replace it with no interruption of the mission. If the satellite fails before the programmed date to ordered it, a satellite production and a launcher request are ordered to replace the failed satellite as soon as possible. This maintenance policy assumes that satellite and launcher productions are not planified but driven by the space segment needs. In case of possibility of satellite stocking, a satellite production is ordered as soon as the previous satellite is launched.

3 MODEL SPECIFICATION

This section describes the models designed for the above presented system. The time parameter is not taken into account here, we only stress the logical behavior. As we showed in section 2, the system could be split into two parts: the space segment and the ground logistic support segment. The global Petri

net of figure 3 follows this decomposition. The sub-model describing the behavior of the space segment. The associated models is presented in sections
3.2. The ground logistic support sub-model describes the management of the
production and launch campaign processes. They are executed in sequence.
Communication between these two sub-models is based on place sharing. Production requests, if marked, allows the start of the production of a satellite and,
concurrently, the order of a launcher. Launch requests, if marked and if a satellite and a launcher are available, allows the launch of a satellite. If this launch
fails (Failure transition), a production is automatically ordered and another
launch request is sent. If launch succeeds, a token is put in Launch indication
which indicates that a satellite is available on orbit. This communication is
shown in figure 2.

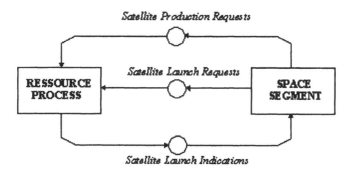

Figure 2. Communication between space segment and ground resources

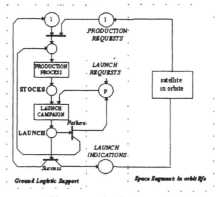

Figure 3. Global model

Production and launch processes are only represented here as transitions (Production Process and Launch Campaign transitions), but, they could be complex processes. Initially no satellite has been produced, launched, or deployed so the initial marking is such that a satellite production and a launcher are required.

4   MODEL EVALUATION

4.1   The choice of Stochastic and Time Petri Net simulation

Ordinary Petri nets are an asynchronous model into which time is not explicitly incorporated. A transition fires from the moment it has been validated, but no hypothesis has been made as to the firing time. These models cannot be used for quantitative analyses, and it is thus necessary to describe a way in which time may be precisely represented.

Several temporal extensions of the basic model have been developed, associating time with the net places, arcs or transitions. A highly general extension was proposed in (Ref. 3), based on the association of time intervals with transitions. An interval is referenced in relation to a transition validation date, and characterizes the set of possible dates for firing this transition. The Petri net can thus be seen as a set of temporal constraints, coordinated by the net structure, and conditioning the net change over time. However, this concept of time intervals, as associated with transitions, is not sufficient. It is advisable, in fact, to specify the probability of the transition firing within a given temporal window. To this end, distribution functions are associated with these time intervals, thus characterizing the random variable for "transition firing time". This is the principle on which Stochastic Timed Petri Net (Ref. 2) is based.

In the case where the time interval is $[0,+\infty[$ and the only possible distribution associated with a transition is the exponential distribution, the model in question is that of Markovian Stochastic Petri Nets (Ref. 4). The change of net marking over time is thus characterized by a Markov process. Finally, if we allow, in addition, immediate Dirac functions, the model obtained is that of GSPN (Ref. 1) which thus allows for synchronizations to be made.

While evaluation methods based on graph generation, like Markov graph (Ref. 1), or probabilized state graph (Ref. 2) are effective because they are formal and analytic, in our specific case, Markov graphs don't enable the inclusion of all types of temporization, while probabilized state graphs supplies only the mean values without their associated standard deviations, which is a significant restriction.

Thus the Monte-Carlo simulation of STPN is used to provide results in both the transient and stationary periods. The simulation is based on multiple random evolution of nets conditioned by the time intervals and associated distribution functions over a given observation period.

The results of the various simulations are averaged and it thus becomes possible to obtain the mean period of sojourn time in the various markings, the average number of transition firings, etc.

It is then easy to deduce the dependability magnitudes in which we are interested.

The simulation tool used for such system evaluations is MISS-RdP (Interactive Modeling and System Simulation using Petri Nets) (Ref. 5) which has been developed by IXI corp. in a partnership which includes academic research (CNRS-LAAS), industrial companies (aeronautics, space and nuclear companies) and the French Space Agency (CNES). It supports STPN model simulation and provides results in standard spreadsheet formats. The collaboration still goes on and a new version for colored STPN is developed and is in evolution to integrate partners needs.

4.2 Simulation parameters

The space system have been modeled by Petri nets in a symbolic way. Hence, a very wide sensitivity analysis over input parameters is allowed without designing new models. We do not proceed it, but illustrate over examples the results we can expect and how they are interesting to analyze and set up the system.

4.3 Results

In order to identify that might achieve the mission availability objectives under the program cost constraints, the study should provide for the entire mission:

- ■   The instantaneous availability during mission time

- ■   The mean availability over the mission time

- ■   The satellite reliability and lifetime

- ■   The satellite production and stock policy

- ■   The satellite cumulated storage time

- ■   The number of satellites launched

- ■   The number of unsuccessful launches

- ■   ...

Most of those outputs are used to evaluate the total cost of recurrent program. Other interesting outputs for the project are provided as the mean number of satellites which could reach half of the lifetime or the lifetime,....

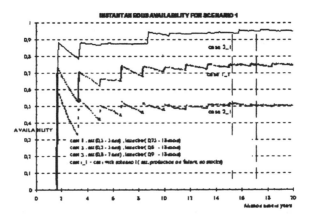

Figure 4. Space segment instantaneous availability

Figure 4 displays the space segment availability for a nominal mission.. The calculus of confidence interval comes from direct application of central limit theorem to the random availability variable (Ref. 6). Figure 6 shows, for 10 000 simulations, the confidence curves between which 95% of experimental values of nominal mission availability should be located. The availability results are accurate enough to allow correct interpretations.

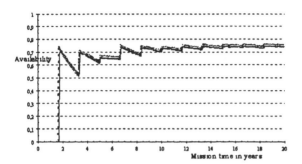

Figure 5. Confidence curves of nominal
mission availability at 95%

In order to evaluate the mission needs, it is interesting to display, at a specific mission date, the number of satellites and launchers required and the associated probabilities. Figures 6 displays the number of satellites ordered at 14 years and the associated probabilities. For example, there is a probability of about 0,8 that 5 satellites or less, and a probability of 0,37 that 5 satellites exactly, have been launched during the first mission time: [0, 14 years].

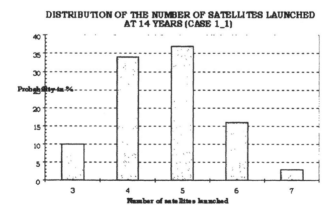

Figure 6. Number of satellites ordered at
14 years and associated probabilities

These are important inputs for costs & risks analysis which underline the necessity to predict launches and satellites failures. Moreover it helps to predict and plan the productions. Costs evaluations are not presented here but the estimations were done in each case studied. The part of each partner of the program (launcher, satellite designer,..) and the total cost were evaluated to compare the different options.

5 CONCLUSIONS

This paper has presented the approach for availability evaluation of space systems developed in the French Space Agency. It has been led in other various cases to analyze commercial or scientific space projects like BIMILSAT (communication), TAOS, STARSYS (localization). From the analyst's view point, Petri net modeling and simulation of such systems, while it requires a non negligible effort of learning and training, allows us to surmount the limitations of traditional approaches. The easy to understand graphical representation of Petri nets improves the dialog between project protagonists who can share the same system view in an unambiguous way. Moreover, Petri net modeling is flexible enough to allow wide sensitivity analyses without designing each time new models. Results of such studies overcome the simple verification of availability objectives. Actually, they help to define optimal deployment and maintenance strategies. Then, guidelines can be provided to project management in order to plan satellite production, negotiate with launcher companies, and estimate global costs.

6 REFERENCES

[1] M. Ajmone Marsan, G. Balbo and G. Conte, "A class of Generalized Stochastic Petri Nets for the Performance Evaluation of Multiprocessor Systems", ACM Trans. on Computer Systems, 2/2, 1984 May, P 93-122.

[2] Y. Atamna, G. Juanole, "Dealing with Arbitrary Time Distribution with the Stochastic Timed Petri Net Model, Applications to Queuing Systems", International Workshop on Petri Nets and Performance Models, Melbourne, Australia, 1991 December.

[3] I. A. Merlin and D. J. Farber, "Recoverability of Communication Protocols - Implications of a Theorical Study", IEEE Trans. on Comm., COM-24 (9), 1976 September, P 1036-1043.

[4] G. Florin, S. Natkin, "Evaluation des Performances d'un Protocole de Communication à l'aide des Réseaux de Petri et des Processus Stochastiques", Journées AFCET Multi-Ordinateurs, multiprocesseurs en temps réel, CNRS, Paris, 1978 Mai.

[5] IXI, MISS-RdP version 4.0, Manuel de l'utilisateur 1994.

[6] F. Moyse, "Précision des résultats de simulation du logiciel MISS-RdP", Rapport de stage de DEA de Mathématiques Appliquées, Ecole Nationale Supérieure de l'Aéronautique et de l'Espace & Université Paul Sabatier, 1995 September.

RESULTS OF LOW-COST PROPULSION SYSTEM RESEARCH FOR SMALL SATELLITE APPLICATION

J. J. Sellers*, T.J. Lawrence*, M. Paul**

United States Air Force
CSER, University of Surrey, Guildford, GU2 5XH, UK
Phone: (44) 1483 300 800 x3411, FAX (44) 1483 259 503
e-mail: t.lawrence@ee.surrey.ac.uk
*** Surrey Satellite Technology Ltd*
Guildford, UK

ABSTRACT

The paper summarizes on-going research into low-cost propulsion system options for small satellite attitude and orbit control. Research into the primary cost drivers for propulsion systems is discussed along with implications for practical, cost-effective designs. Results of hybrid rocket experiments are highlighted. Applications for this technology on future low-cost missions is examined. Other technology options are also reviewed including cold-gas thrusters, resistojets and low-thrust bi-propellant engines. The propulsion system for the forthcoming UoSAT-12 minisatellite system is described in detail along with on-orbit capability and operational modes. Future propulsion research work is summarized.

1 INTRODUCTION

The current catch-phrase of the aerospace industry "faster, better, cheaper" represents political and economic necessity as much as good engineering practice. While industry as a whole struggles to fully define this "new" strategy in the wake of the post-cold war draw down, the University of Surrey and its commercial arm Surrey Satellite Technology Ltd. (SSTL), as well as others in the so- called amateur satellite community, have been quietly implementing this philosophy all along. Since 1981, University of Surrey satellites (UoSATs) have shown that small, reliable satellites can be built and operated at costs far less than one would find in the mainstream aerospace industry. The basic UoSAT design, evolved over more than ten missions, has proven itself as a reliable cost-effective platform for quick access to low Earth orbit. So far, all UoSAT spacecraft have been in the microsatellite range ($\sim$ 50 kg), designed to operate in the relatively benign environment of low-Earth orbit (LEO). As of this writing, the SSTL/UoSAT team have logged nearly 50 orbit-years of operational experience.

The success of small satellite missions depends on low-cost launch opportunities. So far, the majority of UoSAT missions have been on *Ariane* launchers attached to the *Ariane* Structure for Secondary Payloads (ASAP) ring and deployed into LEO. However, a review of *Ariane* launch manifests into the foreseeable future at the outset of our research revealed the majority of opportunities for secondary payloads would be into geosynchronous transfer orbit (GTO). This is a highly elliptical orbit 200 x 36,000 km altitude with an inclination of 7^0. GTO offers a variety of mission opportunities. OSCAR Phase-3 missions, for example, have used GTO as starting point for Molniya communications missions. Other mission opportunities that could exploit low-cost launches into GTO include:

- *Small geosynchronous (GEO) communications*: Currently, developing countries must lease transponders on large, expensive commercial satellites. The possibility exists for them to purchase their own small, dedicated satellite at a competitive price.

- *Meteorological monitoring*: Microsatellites have demonstrated their utility for localised weather monitoring from LEO. Small satellites beginning in GTO could be used as low-cost weather monitoring platforms with the higher altitude providing more global access.

- *Geomagnetic data collection*: Because spacecraft in GTO travel though the entire depth of the Van Allen radiation belts twice daily, they offer a unique vantage point from which to monitor important phenomena in the space environment such as solar wind and magnetic field interactions, galactic cosmic rays and solar flares.

- *Ground-based astronomy calibration*: Ground-based optical astronomy is handicapped by the dynamic nature of Earth's atmosphere which attenuates faint signals. A satellite in very high Earth orbit with a low-power laser of known wavelength could provide the feedback necessary to perform real-time calibration and correction of these signals, greatly enhancing their resolution.

- *Lunar and planetary exploration*: From GTO, the total velocity change (*Delta*V) necessary to go to lunar orbit, Earth approaching asteroids or even other planets is roughly equivalent to that needed to go into GEO. Spacecraft such as the U.S. *Clementine* mission have demonstrated how very good planetary science can be conducted from small, relatively low- cost (~$70M) platforms. The opportunity exists to use GTO as a springboard to explore the solar system with even smaller, and far cheaper (~$10M) satellites.

In addition to these missions specifically related to GTO, other exciting opportunities are emerging for LEO small satellites:

- *Micro-LEO constellations*: Constellations of two or more store-and-forward communication satellites to support world-wide paging, data collection from geographically remote scientific or industrial facilities, disaster relief and other services.

- *"Personal" Remote Sensing*: Developing countries currently depend on large, expensive remote sensing platforms such as SPOT or LANDSAT. A dedicated small satellite with nearly the same resolution that also offers on-demand coverage and the ability for the user to exercise far greater control over imaging times, targets, lighting and area re-visits appears feasible at a competitive cost.

- *SAR missions*: Synthetic aperture radar (SAR) offers the ability to pierce through cloud cover to collect images day or night. So far, this technology has been limited to very large, expensive platforms, however, it now appears feasible to deploy a limited but useful SAR capability on a small

LEO satellite.

- *Equatorial belt missions*: All of the LEO missions listed above would provide global coverage from a polar orbit. However, developing counties, especially in the Pacific Rim such as Indonesia, Malaysia, Singapore and the Philippines are increasingly interested in dedicated regional coverage. This could best be provided by satellites operating in very low inclination orbits.

Unfortunately, until recently UoSAT spacecraft (as well as similar satellites built by other Universities and companies) lacked one critical system that would allow them to exploit fully the mission opportunities outlined above:*a propulsion system*. Propulsion systems are a common feature on virtually all larger satellites. However, until now there has been no need for very small, low-cost satellites to have these potentially costly systems. As secondary payloads, they were deployed into stable, useful orbits and natural orbit perturbations (drag, J2, etc.) were acceptable within the context of the relatively modest mission objectives. Over the years, these pioneering small satellite missions have proven that effective communication, remote sensing and space science can be done from a cost-effective platform. As these missions have evolved, various technical challenges in on-board data handling, low-power communication, autonomous operations and low-cost engineering have been met and solved. Now, as mission planners look beyond passive missions in LEO to the bold, new missions described above, all of which require active orbit and attitude control, a new challenge is faced: cost-effective propulsion.
Propulsion systems are needed to perform a variety of tasks essential to active missions in LEO and beyond. These include:

- *Orbit Manoeuvring*: the ability to move from an initial parking orbit to an escape trajectory or insert into a final mission orbit, e.g. changing from GTO to GEO.

- *Orbit Maintenance*: the ability to maintain a specific orbit against drag and other perturbations, or phase the orbit to maintain proper angular separation of a constellation.

- *Attitude Control*: The ability to rotate the spacecraft to reorient sensors or dump momentum, especially beyond LEO where magnetorquing and gravity gradient stabilisation are not viable options.

Obviously, all these capabilities can be found in off-the-shelf systems used throughout the aerospace community. However, current off-the-shelf technology may not be appropriate for cost-effective applications within the context of small satellite missions. Furthermore, the cost of these systems, when procured using standard aerospace practices can be prohibitive. Thus, small satellite mission planners face a dilemma: future missions demand a propulsion capability but the cost of this single system may be prohibitive, keeping the entire mission grounded.

The objective of the research described in this paper is to investigate cost-effective propulsion system options for small satellite application. The following discussion will address a new paradigm developed for understanding propulsion system costs and an innovative technique for parametrically combining the various dimensions of this paradigm to quantify a figure of merit for specific system options and mission scenarios. This technique provides a useful tool for mission and research planning as well as total quality management. From this discussion, hybrid rockets and water resistojets emerge as promising technology options in need of further research. Research programs investigating these technologies at the University of Surrey will be described along with preliminary results. Finally, the near term application for this propulsion research will be addressed by describing the system to be deployed on the forthcoming UoSAT-12 minisatellite mission.

2 DISCUSSION

2.1 Cost Paradigm

At the outset of the research into the cost issues of propulsion systems, it immediately became obvious that the broader issues of spacecraft hardware costs in general must first be addressed before propulsion system costs specifically could be fully understood. By first isolating and explaining the fundamental cost drivers of these traditionally expensive components, a credible strategy could be formulated for reducing the costs of propulsion hardware specifically. To that end, specific spacecraft hardware cost drivers which occur during each phase of a mission were identified. These mission phases are:

1. Mission Definition

2. Mission Design

3. Hardware Acquisition

The purpose was to provide a useful context for understanding the process of selecting and flying space hardware in general which could then be applied to propulsion systems. The results of this effort are published in a dedicated chapter of*Reducing Space Mission Costs* edited by Dr. Jim Wertz [18].
From this research, a new paradigm for understanding total propulsion system cost emerged. Traditionally, the approach taken to describe propulsion cost has been to isolate a single descriptive parameter of the technology, one that determines what was perceived to be the most important premium on a satellite: mass. The propellant mass used by a given system is determined by its specific impulse, *Isp*. Specific impulse is similar in concept to the "miles per gallon" rating used to compare automobile fuel efficiency. However, while mass is certainly one important descriptive dimension of system cost it is not the only one. In fact, the evidence presented from [2] clearly indicates that by focusing solely on mass, true cost reduction may not be achieved. It is even possible that the overall cost is *increased* due to the increased system complexity needed to achieve the higher mass efficiency.
If mass is not the only dimension, what else is there to consider? The most obvious that springs to mind is the bottom line price paid for the hardware. In some situations, price can be the most important dimension, especially for low-budget missions such as those flown by small, University- sponsored satellites. For these missions, if the price exceeds a certain threshold limit, the mission simply will not get off the ground.
But focusing too closely price alone may cause you to miss more important issues. For example, a given system option may appear to be a bargain in terms of dollars, but ensuing logistics or operating costs may far exceed other, seemingly more expensive options. Therefore, as part of the research, we set out to define all the dimensions that encompass total propulsion system cost. In addition to mass and price, there are three other aspects of performance to consider: volume, Total elapse thrust time (to complete all *Delta*V), and power consumed. Finally, there are other less obvious opportunity costs to consider as well. Collectively, these as referred to as *mission costs* as they depend on the

technology used and the mission environment. These are: technical risk, safety, logistics and integration. Thus, the nine dimensional cost paradigm includes:

1. Propellant mass

2. Propellant volume

3. Total elapsed thrust time (to complete desired *Delta* V)

4. Power required

5. System price

6. Technical risk (to the program)

7. Safety (to deal with inherent personal risk)

8. Integration

9. Logistics

Figure 1 illustrates how each phase of a mission drives the specific cost dimensions. Using this new paradigm, the real cost of system options could then be assessed. This will be addressed in the next section.

2.2 Assessing System Options

Armed with a new paradigm for understanding propulsion system costs, all realistic near-term options for small satellite applications were considered. These options are listed in Table 1 along with important performance parameters. The chemical systems identified included traditional solid and liquid systems as well as hybrid. For electric systems, the study showed that resistojets and pulsed plasma thrusters (PPT) look the most promising for small satellites due to their low power requirement (50 - 500 W continuous power). Ion systems have been designed for low power ($\sim$ 440 W), but are very expensive ($\sim$ 1.5 million for each thruster). Microwave thrusters can also function at low power, but still are in the theoretical stage. The other systems, have too high of a power requirement for the UoSAT platform. It was decided due to the anticipated simple integration requirement for a water resistojet, that it would be worth pursuing (water can go anywhere in the world !). PPT's are being studied at

NASA Lewis and the USAF Phillips Laboratory (with Olin as the contractor)
and will be attractive as soon as a more advanced systems are flown [10].

In addition to understanding performance costs, the price and mission costs
for each option were also characterised. System prices were determined by
designing representative system architectures for each option and then pricing
individual components based on information gained from the UoSAT-12 system
design project (described later in the this paper). Mission costs for each option
were evaluated based on a thorough understanding of each technology applying
engineering judgement. Table 2 lists the relative mission costs for each option.
These and all dimensions were scaled 0 - 100 (with 0 being lowest cost and 100
highest) to allow for parametric combination.

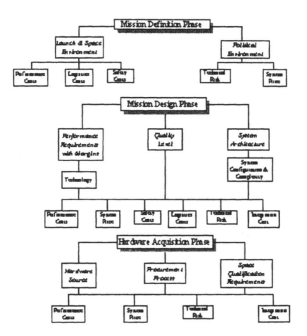

Figure 1: Relationship between mission phase
and cost drivers for propulsion systems.

System	Isp (sec)	Oxidiser/Propellant specific gravity	Fuel specific gravity	O/F	Density Isp	Thrust (N)
Bi-Propellant	290	1.447	0.8788	1.65	337.33	20
Hydrazine mono-propellant	225	1.008			226.80	20
Hybrid	295	1.36	0 93	8	381 60	500
Cold-gas	65	0.23			14 95	0 1
H2O Resistojet	185	1 0			185 00	0.3
Solid (STAR 17-A)	286 7	1 661			476 21	16000
Hydrazine Resistojet	304	1 008			306 43	0.33
PPT	1500	2 16			3240 00	7 0 x 10-4
HTP mono-propellant	150	1.36			204 00	1

able 1: Performance comparisons between various propulsion technologies analysed

Option	Safety Cost Factor	Technical Risk Factor	Integration Factor	Logistics Factor
Bi-Propellant	100	50	80	100
Hydrazine mono-propellant	90	40	70	90
Hybrid	50	100	100	60
Cold-gas	10	10	10	20
H2O Resistojet	10	80	20	20
Solid	20	30	100	80
Hydrazine Resistojet	90	40	40	90
PPT	10	80	80	10
HTP mono-propellant	50	80	70	60

Table 2: Relative mission costs for propulsion system options.

The approach taken to parametrically combine all cost dimensions into a single figure of merit is illustrated in Figure 2. A mission scenario was first defined which determined the mission environment and the total ΔV required. The mission environment determines the weighting applied to each dimension. For example, an experimental mission sponsored by a University will place a higher premium on price than mass while a commercial organisation may take the opposite approach. Each dimension was weighted from 8 - 0 with 8 being the most important and 0 meaning no importance (e.g. a given mission may place no weight to the total time needed to complete a manoeuvre). The mission ΔV was used to determine the performance, price and mission costs for each option. All parameters were then combined using the following relationship:

$$Total\_cost = A.Prop\_Mass + B.Prop\_Vol + C.Time + D.Power +$$

$$E.Logistic + F.integrat + G.safety\_Cost + H.Tech\_Risk + I.Sys\_Price$$

(2.1)

where

A-I = Weightings on each dimension

To illustrate the utility of this method, it was applied to four different mission scenarios.

- *Traditional commercial mission*: 200 m/s ΔV station keeping. Performance costs dominate. (Solid option not considered)

- *Non traditional, SSTL type commercial mission*: 200 m/s ΔV station keeping. System price dominates. (Solid option not considered)

- *Experimental mission*: Low ΔV (20 m/s). System price dominates.

- *High risk Lunar orbit mission*: High ΔV (1600 m/s). Initial geosynchronous transfer orbit puts a premium on time due to total radiation dose effects. (Cold-gas and PPT options not considered)

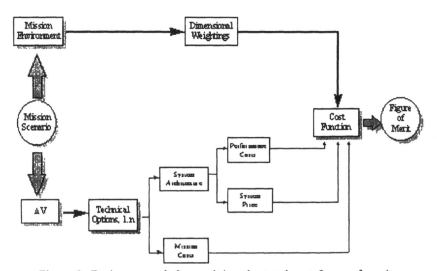

Figure 2: Basic approach for applying the total cost figure of merit.

The weightings applied to each dimension for each scenario is shown in Table 3. Results are shown in Table 4 and Table 5. Total propellant mass for each option and mission is shown along with the total estimated system price and the normalised total cost computed from the cost figure of merit.

It is important to emphasise that the results presented above are not intended to be the final word. The purpose in developing this process was to produce results from which to start discussion, not iron-clad answers intended to end all debate. For example, different schools of thought may take exception with the exact qualitative values assigned to certain technology options or the weightings applied to various dimensions for a given mission scenario. The results may differ depending on the actual assumptions made, although, the process used remains the same. Therefore, the more general conclusions about the utility of the process itself are the most important to consider.

Rank (weighting)	Traditional Commercial Mission	Non-traditional Commercial Mission	Experimental Mission	Lunar Orbit Mission
8	Mass	Price	Price	Time
7	Volume	Integration	Integration	Price
6	Technical Risk	Safety	Logistics	Safety
5	Integration	Logistics	Safety	Logistics
4	Logistics	Technical Risk	Power	Integration
3	Safety	Mass	Volume	Technical Risk
2	Price	Volume	Technical Risk	Volume
1	Power	Power	Mass	Mass
0	Time	Time	Time	Power

Table 3: Weightings applied to cost dimensions for various mission scenarios.

Mission	System	Propellant Mass (kg)	System Price ($)	Cost Figure of Merit
Traditional Commercial	PPT	3 37	$500,000	47 4
	Hydrazine Resistojet	16 22	$229,942	74 3
	H2O Resistojet	26 09	$119,604	77.7
	Hybrid	16 69	$171,701	84 3
	Bi-Propellant	16 97	$176,987	84.9
	Hydrazine mono-propellant	21 66	$132,724	88 8
	HTP mono-propellant	31 77	$122,724	100 0
	Hydrazine Resistojet	1 67	$217,160	3 2
Non-Traditional Commercial	H2O Resistojet	26 09	$119,604	56 2
	PPT	3 37	$500,000	76 1
	HTP mono-propellant	31 77	$122,724	86 3
	Hydrazine Resistojet	16 22	$229,942	90 0
	Hybrid	16 69	$171,701	91 7
	Hydrazine mono-propellant	21 66	$132,724	92 0
	Bi-Propellant	16.97	$176,987	100 0

Table 4: Summary of cost analysis results for traditional vs. non-traditional commercial mission.

Mission	System	Propellant Mass (kg)	System Price ($)	Cost Figure of Merit
Experimental	Cold-gas	7.72	$77,594	30.9
	H2O Resistojet	2.82	$94,040	46.7
	HTP mono-propellant	3.37	$97,160	65.4
	PPT	0.34	$200,000	67.4
	Hydrazine mono-propellant	2.26	$107,160	79.5
	Hybrid	1.72	$171,701	89.2
	Hydrazine Resistojet	1.67	$217,160	97.9
	Bi-Propellant	1.75	$176,987	100.0
Lunar orbit	Hybrid	106.18	$248,393	49.0
	HTP mono-propellant	165.72	$237,762	59.1
	Hydrazine mono-propellant	128.90	$260,544	59.8
	Bi-Propellant	107.54	$279,243	64.0
	Solid (2 x STAR 17-A)	108.46	$1,420,000	68.5
	H2O Resistojet	148.98	$272,988	69.6
	Hydrazine Resistojet	103.80	$332,198	100.0

Table 5: Summary of Total System Cost analysis results for an experimental mission and a Lunar orbit mission.

To begin with, this unique method for comparing system options provides a versatile tool for mission planners that allows them to quickly quantify and compare all available technologies and assess their relative total mission costs. Thus, for the first time, complex system information can be easily quantified. Low-cost satellite engineering at the University of Surrey and elsewhere has epitomised the virtue of applying appropriate technology to a given problem. The appropriateness of a technology is judged by taking a wider view that encompasses more than simply price or performance. Until now, engineers typically relied on completely subjective engineering judgement or "gut feeling" in order take into account such indirect cost factors as integration and safety. The total cost figure of merit process now provides a quantifiable means of making those important engineering decisions.

Furthermore, because the process results in a quantifiable parameter, it can serve as a useful total quality planning tool. By quantifying the starting point for various options, this technique can provide important indications of where best to invest in improvement and enables any incremental improvements to be measured. In this way, the controversial results reported above may help to spark debate and force a re-examination of research priorities for small satellite propulsion. For example, in deciding where best to invest money in a PPT development program, this process (as evidenced by the results above) would indicate that more effort should be aimed at lowering the price and integration complexity of the thruster rather than on increasing its delivered *Isp*. In doing this, it would clearly offer a better overall cost-effective solutions to competing options.

The most immediate application for this method as a research planning tool is at the University of Surrey. These results indicate that three propulsion technologies offer real benefit for future mission scenarios:

- *Hybrid rockets*: for future high *Delta*V options such as the Lunar mission (with the HTP mono-propellant system as a necessary bi-product of such research).

- *H2O resistojet*: for commercial applications for the minisatellite bus for station keeping requirements in LEO or GEO.

- *Cold-gas*: for near-term experimental missions to develop basic orbit control techniques.

These research areas will be addressed in the following sections.

2.3 Hybrid Rocket Research

Based on the analysis presented in the last section, hybrid rockets emerged as a promising technology in need of further research. Hybrid rockets offer an inherently safe option that use a liquid oxidiser and a solid fuel. In operation, they cannot explode. The following subsections describes the research. Additional background on the program can be found in [12] [13].

Research Objectives

Beginning in April 1994, supported by UoSAT/SSTL, we undertook this ambitious hybrid rocket research programme. The goals of the program were established as follows:

1. *Proof-of-concept*: demonstrate the accessibility of hybrid rocket technology for continued research, development and exploitation for low-cost satellite upper stages. In the process, identify and solve the critical engineering

problems of the technology.

2. *Performance characterisation*: recognising that the actual performance of a given hybrid propellant combination depends on empirical data, establish through experimental investigation, the regression rate characteristics for a proto-type motor and use this data as a basis for preliminary upper stage design calculations.

3. *Total cost assessment*: based on the experience gained through hands-on hybrid rocket work, fully assess the system price and mission costs for future hybrid upper stage applications.

With these objectives in mind, a proto-type hybrid motor was designed, built and tested using 85% high test hydrogen peroxide oxidiser (HTP) and polythene fuel.

Results & Conclusions

The primary objectives of the hybrid research program have already been met. To begin with, the concept has been proven. Hybrids represent a readily assessable technology allowing full-scale research and development in a budget-constrained, University environment. The program demonstrated rapid results (first successful test less than 7 months from project go-ahead) with minimum cost (< \$20,000) and addressed and solved a number of fundamental engineering problems, most notably catalyst pack technology. 55 catalyst pack tests were completed using 8 different catalyst types. 9 successful hybrid tests were completed.

Experimental results allowed the complete characterisation of hybrid performance. The proto-type hybrid motor was used to fully assess the PE/HTP combination and publish the first-ever regression rate relationship applied specifically to small satellite upper stages. This data is shown in Figure 3. Using this data, a hybrid upper stage design process was developed and a preliminary design for minisatellite motor was completed. Performance parameters for this motor are shown in Table 6.

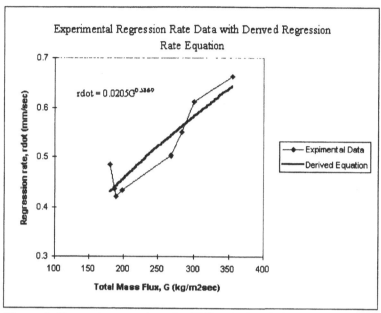

Figure 3: Regression rate data for 85% HTP/PE hybrid combination.

Parameter	Value
Initial port radius (m)	0.008
Initial L/D	25
Ave Isp (sec)*	299.6
Ave O/F	8.4
Total propellant mass (kg)	16.5
Fuel mass (kg)	1.8
Oxidiser flow rate (kg/s)	0.123
HTP volume (litre)	10.8
Ave Thrust (N)	404
Thrust time (sec)	120
Total impulse (Ns)	48,480
Throat diameter (m)	0.0062
Expansion Ratio	150:1
Nozzle length (m)	0.268
Motor length (m)**	0.25

Table 6: Results of spacecraft hybrid motor design exercise.
Performance is assumed to be 95%. Port geometry is assumed to be "double-D".

Finally, the total cost of hybrids with respect to the cost paradigm were assessed. The performance costs for 200 m/s *Delta*V motor were defined above.

Development price for the motor described earlier were estimated to be $100,000 with total system cost ~$170,000.

Obviously, the first flight of any new propulsion technology caries technical risk to the mission. Fortunately, the inherent nature of hybrids makes the chance of a catastrophic failure extremely low. Once hybrid technology has overcome the stigma of being untried in space, these inherent features would make its overall technical risk roughly equivalent to mono-propellant technology.

The combined thermal control and ADCS aspects of hybrid motor integration costs would be roughly the same as those discussed for solid motors. In addition, hybrids have a significant overall integration advantage over solids in that the motor can be fully integrated within the spacecraft prior to shipment. Launch site preparation would require only the loading of oxidiser. Figure 4 gives a cut-away view of the hybrid motor described earlier installed in the minisatellite structure. The mechanical integration complexity for a hybrid would be similar to a mono-propellant system in terms of overall requirements for support systems (tanks, valves, etc.).

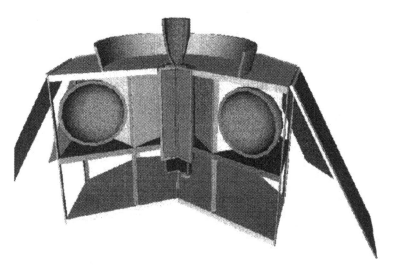

Figure 4: Cut-away diagram showing possible configuration of
a hybrid motor and support tanks within the minisatellite structure.

Safety issues associated with small satellite applications for hybrids arise from two sources:

- Storage and handling of high pressure gas

- Storage and handling of HTP

The first safety issue is not unique to hybrids and must be addressed as part of cold-gas or other liquid propellant options (bi-propellant, mono-propellant or resistojet). As discussed in Chapter 4, procedures and regulations governing high pressure gasses are well established. References such as [MS1522A] specify design criteria for tanks and lines to ensure safe operation.

The second issue is the most important to address. A fair assessment of the safety aspects of HTP must be done in context with other propellant options such as hydrazine, MMH or MON. Both [1] and [3] provide extensive background on HTP safety requirements.

While HTP can cause skin irritation, [3] classifies it as non-toxic in sharp contrast to other liquid propellants. This greatly alleviates demands on the necessary safety infrastructure as "respiratory protection is ordinarily not required" [1] in sharp contrast to hypergolics which, as Chapter 4 describes, require the use of full SCAPE suits. For HTP handling, much less expensive and complex vinyl-coated trousers, coats and hoods with Plexiglas face protection are sufficient. Ordinary rubber gloves (purchased from TESCO) offer adequate protection for equipment handling.

The biggest potential impact of safety considerations on mission costs relates to logistics. HTP must either be delivered to the launch site by the supplier (e.g. Air Liquide) or directly to the satellite manufacturer for shipment as part of the overall launch campaign. According to [1], HTP is authorised for transport aboard military aircraft when packaged in accordance with DOT regulations which defines it as an oxidiser under UN2015. Unfortunately, it cannot be carried on commercial aircraft in any quantity. However, discussions with supplier Air Liquide [15] indicates that the rules governing HTP ground transport are the same that apply to 70% hydrogen peroxide which is used world-wide in the pulp and paper industry. Therefore, ground transportation of even very large quantities virtually any where in the world would be relatively easy to arrange given sufficient delivery notice.

2.4   Water Resistojet Research

This section will describe the water resistojet research at the University of Surrey. Background on the technology will first be discussed. Research objectives will then be reviewed followed by a discussion of preliminary results.

Background

A resistojet can be classified as an electrothermal thruster in that electrical energy is used to directly heat a working fluid. The resulting hot gas is then expanded through a converging-diverging nozzle to achieve high exhaust velocities. As with chemical rockets (which produce heat stored in chemical bonds), the same concerns exist for the relative energy in kinetic vs internal energy, as well as for the loss of energy due to heat transfer and radiation. The primary difference is that for resistojets, the electrically heated channel wall has a higher temperature than the flow. Thus, the performance is limited by the channel wall temperature. However, the advantage of a resistojets is that any working fluid can be used as a propellant. Figure 8 shows the specific impulse, *Isp*, and density specific impulse for various working fluids for a gas chamber temperature, *Tc*, of 1000 K.

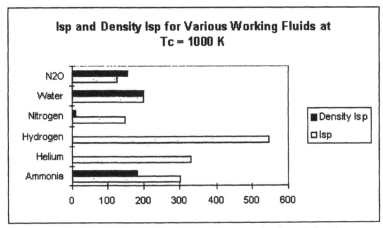

Figure 5: Comparisons of specific impulse (Isp) and density specific impulse for various resistojet propellant options.

The advantage of using water as the working fluid was illustrated in the analysis presented above. Not only does it offer high density and low mass in terms of performance costs, but its inherently safe nature make potential technology development costs low as well.

Research Objectives

Beginning in November, supported by UoSAT/SSTL, we started a development effort studying electric propulsion options for small satellites. Based upon the results discussed above, and a preliminary feasibility study, water resistojets looked very attractive for stationkeeping missions. The goals of the programme were established as follows:

1) *Proof-of-concept*: Demonstrate the accessibility of water resistojet technology for continued research, development and exploitation for low cost satellite stationkeeping missions. In the process, identify and solve the critical engineering problems of the technology. This goal was broken down into the following tasks:

   - Design a proof-of-concept thruster (called the *Mark-I*),
   - Establish test infrastructure,
   - Investigate heat transfer trade-offs.

2) *Performance characterisation*: Unlike performance for chemical rockets which can be readily predicted from combustion thermochemistry, the theoretical performance of a water resistojet must rely on less analytical approximations of heat transfer efficiency. This can become a complicated problem as the heat-transfer correlations for the various transitions of water-steam boiling and two-phase flow are difficult to model- thermodynamic properties are very dependent on the steam/liquid mixture. The high temperatures and low thrust for electrothermal thrusters has traditionally led to a requirement for highly sophisticated (and very expensive) thrust stands to get accurate performance characterisation. One of the primary research goals during this phase will be to develop an analytic technique to acutely model resistojet performance based upon thermodynamic heat transfer efficiencies. These predictions will be compared to experimental results. Once a credible analytic approach is established, the thruster will then be tested on a state-of-the-art thrust stand to further validate the technique. It is envisioned that such an analytic performance

prediction technique will enhance the state of the art for these types of thrusters by eventually reducing or eliminating the need for complex and expensive thrust measurement equipment. The end result will be a low-cost approach to electrothermal rocket testing. This objective will also consist of three tasks:

- Based on experience gained during Phase-1, design a more efficient experimental thruster (called the *Mark-II*)
- Define performance prediction trade-offs and heat transfer correlations
- Collect experimental data over a wide range of operating regimes
- Address satellite integration and operations issues

3) *On-Orbit Demonstration*: The final goal of this research will be an on-orbit demonstration of the technology. This will represent a true "first" as no water resistojet has ever been tested in space. This experiment will be conducted on the UoSAT-12 mission (described below). Care must be taken to ensure the experimental thruster firings are compatible with the spacecraft's duty cycle. Simulation of the operations concept will be incorporated early on in the research programme to insure the optimum solution is obtained. The duty cycle will be optimised for the UoSAT-12 mission but will be designed to demonstrate the operational flexibility of the technology to meet the requirements for future station keeping missions. A proto-flight thruster (called the *Mark-III*) based on the following performance parameters is envisioned for UoSAT-12:

- Thrust = 0.05 - 0.5 N
- Specific impulse = 180 - 220 sec
- Power = 100 - 560 W
- Duration = > 250 minutes total operation.

Program Status

The program is currently in the Proof-of-Concept Phase. The *Mark-I* thruster has been designed and fabricated. For the *Mark-I* design, it was understood that there were many ways the thrust chamber could be configured to give efficient heat transfer to the propellant (a tube wrapped in a electric coil, directly exposing the propellant to an electric coil, a heater surrounded by sintered material or a heater surrounded by a packed bed of heat transfer material). Such designs are discussed in [7] and [8]. We decided to pursue the packed bed approach as it provides high surface area and therefore the potential for high heat transfer. An additional advantage of a packed bed is the relatively high pressure drop created which allows for very long propellant stay time, further increasing heat transfer efficiency. Predicted results for various heat transfer material is shown in Figure 6.

The *Mark-I* thrust chamber is 30 cm by 120 cm with a 10 by 110 cm commercial cartridge heater installed in the centre. Around the heater, the chamber is packed with a heat transfer material (leading candidates are stainless steel, boron carbide, and silicon carbide) in the form of pellets varying from 300 - 700 μm in diameter. Water flow rate varies from 0.05 to .1 g/sec at an inlet pressure of 10 bar. As it enters the chamber, the water passes through a 2 mm sintered disk which keeps the heat transfer material from interacting with the injector and also provides a pressure drop to decouple the inlet pressure from the chamber pressure (otherwise flow oscillations can regulate the inlet flow). The water then flows across the bed, is heated, and passed out through the .5 mm throat diameter nozzle as super-heated steam. The instrumentation in the thrust chamber consists of two pressure gauges and 12 thermocouples. A cut-away drawing of the thrust chamber is shown in Figure 7.

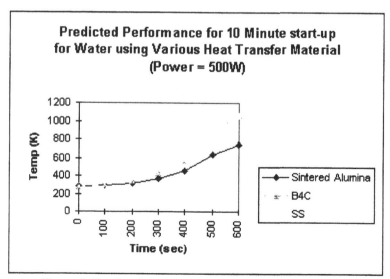

Figure 6: Predicted performance for water resistojet with

various be materials (SS = stainless steel, B4C = boron carbide)

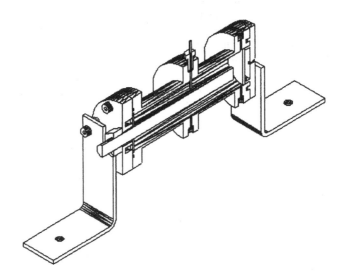

Figure 7: Cut-away diagram of experimental water resistojet.

Preliminary tests using stainless steel as the heat transfer material have been conducted for up to 6 hours of thruster operation expanding to atmosphere. As the program is still in the proof-of-concept phase, actual thruster performance data has not yet been produced. However, the test infrastructure has been validated and the *Mark-I* design has proved useful for understanding heat transfer trade- offs. Initial results indicate reliable heat transfer within the bed with temperatures > 600 K achievable, which is consistent with earlier results [7]. The *Mark-II* thruster is currently being designed for full-scale performance characterisation beginning in July 1996. We are looking at techniques for improving heat transfer efficiency (e.g. insulation, pre-heaters, greater stay time, etc.). This program is on a fast-track to produce a proto-flight thruster for UoSAT-12 by December 1996.

2.5 Uosat-12 System

Concurrent with this research, engineers at SSTL/UoSAT were designing a flexible, multi-mission minisatellite to position themselves to exploit the emerging opportunities discussed above. With an approximate mass of 250 kg, the minisatellite structural design builds on the modular approach used in the UoSAT microsatellites in a way that allows maximum re-use of subsystems between the two platforms. The minisatellite structure starts with a honeycomb attach frame on which three stacks of module boxes are arranged in a triangle. Solar panels of the same width as those used on the microsatellites are arranged around the sides to get a total of nine sides. By extending the height of the panels, an equipment or payload bay is formed at the top of the module box stack. A diagram of the minisatellite is shown in Figure 8. As this is written, the first flight of this new satellite bus, dubbed UoSAT-12, is in critical design for a launch in Mid-1997. The technical objectives for the minisatellite mission strike a compromise between all the features a flexible minisatellite bus would have and what can be achieved within the available budget and time scale. The following technical objectives have been defined for the UoSAT-12 mission:

- Demonstrate a commercially viable minisatellite bus with industry standard support systems

    - 28 VDC power bus
    - 1 MBPS S-band down-link

- Demonstrate that enhanced core microsatellite technologies can be used in a minisatellite:

    - Intel 386-based on board computers (OBC)
    - Low-rate VHF/UHF data links
    - Distributed TT&C via control area network (CAN)

- Demonstrate major new subsystems:

    - Enhanced attitude determination and control capability
    - Propulsion system capability with orbit maintenance and attitude control

- Enhance existing UoSAT payloads using resources of the minisatellite to provide operational demonstration of:

    - High-resolution (< 30 m) multi-spectral visible imaging

– Store-and-forward communications to small terminals

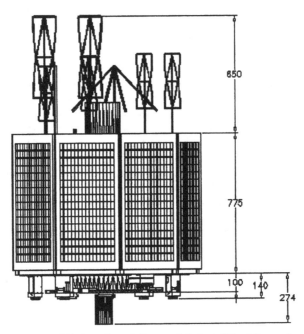

Figure 8: Diagram of University of Surrey Minisatellite (dimensions in mm).

Figure 9 shows the schematic of the propulsion system layout for UoSAT-12. Note as this is an experimental mission designed to gain experience in propulsion system integration and operations, the most cost-effective option (as indicated earlier) was a cold-gas system. A separate water resistojet experiment is also planned. Table 7 summarises the total performance available.

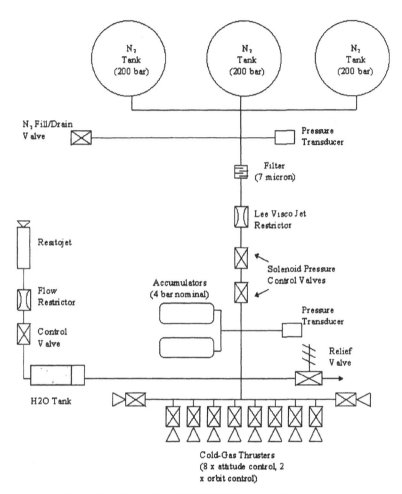

Figure 9: Schematic of UoSAT-12 propulsion system.

Performance Parameter	Value
Mass N2	7 1 kg
Total Pulses	4 384 x 105 (0.1 sec each)
Total Impulse (cold gas)	4.389 x 103 Nsec
Total Angular Impulse (cold gas)	2.085 x 103 Nmsec
DV available (cold gas)	17 8 m/s
Mass H2O	1 5 kg
Total Impulse (resistojet)	1.5 x 103 Nsec

Table 7: Summary of performance parameters for UoSAT-12 propulsion system.

3 CONCLUSIONS

The most cost-effective propulsions system can only be found by weighing all options along the nine dimensions of the total cost paradigm within the context of a given mission. For very low-cost, logistically constrained missions, unconventional options such as hybrids and water resistojets offer many unique advantages over current off-the shelf options. Future research will focus on demonstrating these technologies in orbit.

4 REFERENCES

[1] Chemical Propulsion Information Agency Publication 394, Vol. III, Hazards of Chemical Rockets and Propellants, Johns Hopkins University, Applied Physics Laboratory, Laurel, Maryland, September 1984.

[2] Dean, E., Unal, R., "Designing for Cost", Transactions of the American Association of Cost Engineers, pp. D.4.1 - D.4.6, Seattle, Washington, 23- 26 June, 1991.

[3] Hydrogen Peroxide Handbook, Chemical and Material Sciences Department, Research Division, Rocketdyne: a division of North American Aviation, Inc., Canoga Park, California, Technical Report AFRPL-TR-67-144, July 1967.

[4] Humble, R., Henry, G.N., Larson, W.J., "Space Propulsion Analysis and Design", McGraw-Hill, Inc., College Custom Series, 1995.

[5] Manzella, D.H., Carney, L.M., "Investigation of a Liquid-Fed Water Resisto-jet Plume", NASA Technical Memorandum 102310, AIAA-89-2840, Prepared for the 25th Joint Propulsion Conference, Monterey, California, July 10-12, 1989.

[6] Moore, G. E., Berman, K., "A Solid-Liquid Rocket Propellant System", *Jet Propulsion*, November, 1956.

[7] Morren, W.E., Stone, J.R., "Development of a Liquid-Fed Water Resistojet", AIAA- 88-3288, AIAA/ASME/SAE/ASEE 24th Joint Propulsion Conference, Boston, Massachusetts, 11 - 13 July 1988.

[8] Morren, W.E., "Gravity Sensitivity of a Resistojet Water Vaporiser", NASA Technical Memorandum 106220, AIAA-93-2402, 29th Joint Propulsion Conference, Monterey, California, June 28-30, 1993.

[9] Morren, W.E., MacRae, G.S., "Preliminary Endurance Tests of Water Va-porizers for Resistojet Applications", AIAA-93-2403, 29th Joint Propulsion Conference, Monterey, California, June 28-30, 1993.

[10] Myers, R.M., Oleson, S.R., "Small Satellite Propulsion Options", AIAA 94-2997, 30th Joint Propulsion Conference, Indianapolis, Indiana, June 27-29, 1994.

[11] MIL-STD-1522A Standard General Requirements for Safe Design and Op-eration of Pressurized Missile and Space Systems, United States Air Force, 28 May 1984.

[12] Sellers, J.J., Astore, W.J., Crumpton, K.S., Elliot, C., Giffen, R.B., Larson, W.J. (ed), *Understanding Space: An Introduction to Astronautics*, McGraw-Hill, New York, N.Y., 1994.

[13] Sellers, J.J., Meerman, M., Paul, M., Sweeting, M., "A Low-Cost Propul-sion System Option for Small Satellites", *Journal of the British Interplanetary Society*, Vol. 48., pp. 129-138, March, 1995.

[14] Sellers, J.J., Paul, M., Meerman, M., Wood, R., "Investigation into Low-Cost Propulsion System Options for Small Satellites" Presented at the 9th Annual AIAA/USU Small Satellite Conference, Logan, Utah, September 1995.

[15] Tremblot, A., Air Liquide, private communications, February March 1996.

[16] Wernimont, E.J., Heister, S.D., "Performance Characterisation of Hybrid Rockets Using Hydrogen Peroxide Oxidiser", AIAA-95-3084, 31st AIAA/ASME /SAE/ASEE Joint Propulsion Conference and Exhibit, San Diego, California, 10-12 July 1995.

[17] Wernimont, E.J., Meyer, S.E., "Hydrogen Peroxide Hybrid Rocket Engine Performance Investigation", AIAA 94-3147, 30th Joint Propulsion Conference, Indianapolis, IN, June 1994.

[18] Wertz, J. (editor), "Reducing Space Mission Costs", Kluwer Publishing, to be published 1996.

16

MULTIPLE TECHNOLOGY CHOICES FOR A MIXED ANALOG-DIGITAL SPACE RADIO-ASTRONOMY SPECTROMETER

J-L. Noullet*, L. Ravera**, A. Ferreira*, D. Lagrange**, M. Giard**, M. Torres***

*LESIA - Institut National des Sciences Appliquees
31077 Toulouse, France
**Centre d'Etude Spatiale des Rayonnements, Toulouse, France
***Institut de Radio-Astronomie Millimétrique, Saint-Martin d'Hères, France

ABSTRACT

For analysing the sub-millimeter radiation received by a astronomy satellite, a hybrid analog-digital spectrometer is planned. As a critical part of this instrument, a specific digital signal processing chip set is proposed. A first strategy is based on a GaAs MESFET chip expected to perform 52 giga-operations per second with a 400 MHz clock frequency. While this strategy gave encouraging results and remains the main line of the project, a second strategy based on parallelism on silicon only is under study, and the opportunity of working simultaneously on two opposite approaches is discussed.

1   THE INSTRUMENT

The analysis of sub-millimeter radiation (0.1mm $< \lambda <$ 1mm) coming from "cold media" can help understanding the composition of interstellar clouds, observing the generation of stars and planets and even detecting planets and their natural satellites.

The frequency spectrum of this radiation carries information not only on the physical and chemical composition of the objects, but also on the relative speed of these objects (Doppler effect).

An example of project aimed at exploring this new domain is the FIRST satellite (Far InfraRed and Sub-millimeter Space Telescope) that is to be launched by the European Space Agency around year 2006. (figure 1)

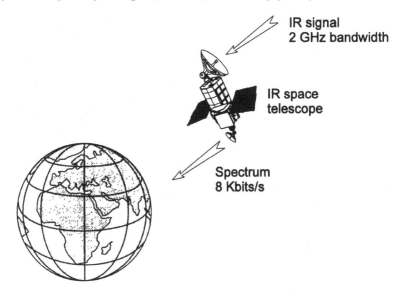

IR signal
2 GHz bandwidth

IR space
telescope

Spectrum
8 Kbits/s

Figure 1   Satellite bearing the spectrometer

In such a satellite, the raw data provided by the sub-millimeter receiver occupies a wide bandwidth (for example 2 GHz) even when the meaningfull spectrum information represents a only few kHz.

The reason of this bandwidth reduction is that the spectral values must be integrated during at least some seconds to obtain something else than pure random noise.

(Notice that the initial signal/noise ratio is one order of magnitude below unity)

It is absolutely necessary to perform this processing (analysis and integration) on board of the satellite, in order to keep at a reasonable level the cost of transmitting the data down to the Earth.

This implies that the instrument will have to comply with very severe constraints of weight, power consumption and reliability.

2 THE MIXTURE OF ANALOG AND DIGITAL

Pure analog solutions exist: for example the acousto-optical spectrometer [1]. It gives fine results, but has the following drawbacks:

- high sensitivity to environment (temperature, vibrations),

- lack of flexibility, specially in terms of spectral resolution,

- accuracy drift problems, like any analog processor.

A pure digital solution with the required bandwidth is not permitted by the present state of the art.

The authors chose to develop an advanced hybrid analog-digital instrument, with the following expected benefits [2]:

- some kind of robustness, easy "space qualification"

- possibility of improving the spectral resolution as it is needed

- stable accuracy

In the proposed approach, the initial bandwidth of some 2 GHz will be first cut into slices by means of analog circuitry, each slice some 180 MHz being down-converted and then sampled at 400MHz and digitized (this is the minimal possible amount of analog processing) (figure 2). Then the digital part of the instrument will compute the integrated autocorrelation function of the signal, easier to compute than the actual spectrum and carrying the same information at the same cost (the spectrum will be computed on Earth by means of the Fourier transform).

One question is what resolution is needed for the analog to digital conversion. Theoretical computation [3] showed that the value of two bits is suitable for the project.

At first glance, one may be surprised reading that two bits only are used to digitize a signal containing much more noise than information. To give a simplified explanation, let us state that the statistical properties of the thermal noise and of the quantification noise allow to reject them after processing thousands of millions of samples, even if both noises are much greater than the signal coming

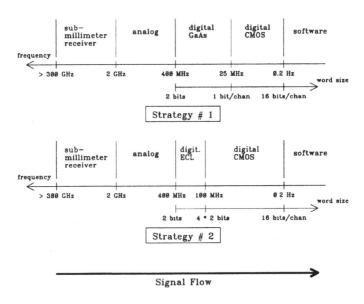

Figure 2   The signal flow in the two strategies

from the interstellar cold media.

In the other hand, only a few microvolts of deterministic noise injected in the analog part may corrupt the measurements, this is why the analog part must be well separated from the digital one.

3   THE BASIC AUTOCORRELATOR, OR THE STRATEGY # 1

The discrete autocorrelation function of a digitized signal can be computed by means of time delays and multiplications. A regular structure is constructed on the basis of a folded delay line: fig. 3.

This structure can be described as a set of similar channels, each channel computing one sample of the autocorrelation function. Channel N gives the product of the signal with its image delayed by N times the sampling period, while channel zero gives the square of the signal.

These products are then integrated by accumulators (one per channel, not represented on fig. 3), and the contents of the accumulators is transmitted to the user at the end of each integration period (a few seconds).

A large accumulation capacity is needed (e.g. 35 bits), since the integration period (e.g. 5 seconds) is very long compared to the sampling period (2.5 ns). In fact, only the most significant bits (e.g. 16 bits) of the accumulated value are meaningful for the user, due to the domination of thermal noise in the incoming signal.

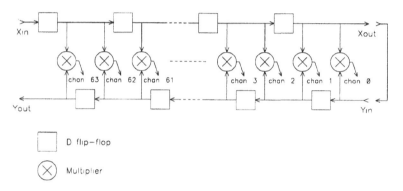

Figure 3   The single folded delay line architecture

Notice that in figure 3, the signal lines carry actually 2 bits of data, and each of the square boxes representing the delay elements contains actually 2 flip-flops.

3.1   GaAs against silicon in strategy # 1

The power consumption is the most critical issue for this system, like for any space equipment.

On the silicon side, the bipolar ECL logic could run at the desired speed, but with a great power consumption and a poor area efficiency.

CMOS is not as fast, even if submicron CMOS circuits have been run at 200 MHz. The problem with ultra high speed CMOS is the power consumption, partly because of the large voltage swing of the logic signals.

A good competitor is the GaAs MESFET technology with DCFL gates (Direct Coupled FET logic). The DCFL gates have a simple architecture, built with enhancement and depletion N-channel JFETs in a manner that recalls the early NMOS logic, even if the electrical behaviour is quite different due to the direct gate-to-source current [4].

The small logic voltage swing and the small supply voltage are the keys of a good speed/power ratio [5].

DCFL is the most compact solution for digital GaAs, the integration density is comparable with static sub-micron CMOS.

A high integration density is wanted not for cost reasons, but mainly because it has an indirect influence on the total power consumption: a poor density would lead to a greater number of interconnected chips, dissipating lots of power in I/O buffers.

Let us remark that below some 60 MHz, CMOS consumes less power than GaAs, and that the difference is very important at low frequencies.

In the autocorrelator, a significant part of the integration task can be performed below 60 MHz. This is why the proposed architecture is composed of a GaAs chip for shifting, multiplications and first stages of integration, down to a bandwidth of 12.5 MHz, (sampling at 25 MHz) and a CMOS chip for the remainder of the integration (figure 2, strategy # 1).

3.2   Mixing synchronous and asynchronous design

The desired integration duration requires an accumulator capacity of 35 bits for each channel.

Considering that at the end of the integration period only the 16 most significant bits will be transmitted to the end user, we note that the 19 less significant bits of the accumulated result do not need to be available in parallel form.

Thus, a solution using an asynchronous cascade counter to perform the integration would be minimal in terms of area and power, and would allow to defer to a CMOS chip some three quarters of the integration task.

The complete solution is complicated by the fact that the actual output of each mutiplier must be first accumulated synchronously, and it is the overflow of this operation that will be chopped and directed to an asynchronous cascade counter (5 stages down to 12.5 MHz bandwidth).

The outputs of the 64 channels of the GaAs chip have to be multiplexed by 4:1 for transmission to the CMOS chip using a reasonable number of I/O pads. This synchronous multiplexing requires the overflows to be re-sampled at 25 MHz, the multiplexed output occupying then a 50 MHz bandwidth.

The timing of these conversions between the synchronous and asynchronous worlds had been very carefully checked, in order to avoid the production of spikes (when chopping synchronous data to obtain an asynchronous clock) and setup-hold timing violations or meta-stable states (when re-sampling the output of an asynchronous counter).

3.3 Full custom against standard cells in strategy # 1

Only a full custom design could take advantage of the regular structure, potentially allowing very short interconnections. A short interconnection length allows to build gates with smaller devices than in the standard cell context, leading to a power saving [6].

The design cost of a full custom layout of the autocorrelator is kept affordable by the use of N similar instances of the "channel cell", with most of the wires connected by abutment. (N = 64 in the first prototype).

The gate metal is used for short wires inside the channel cell, reducing the number of vias and allowing a much more dense layout than with standard cells. It has been checked that the resistance of these wires (max. 150 Ohms) induces a negligible delay (less than 5 ps).

A long shift register is the most sensitive circuit when considering the clock skew hazard.

To minimize the clock skew, a low-impedance planar grid clock net was made, driven by 16 buffers evenly spaced on the chip core. Such a precaution is possible only in a full custom context.

4 THE PARALLEL APPROACH, OR THE STRATEGY # 2

The idea: dividing by a factor K the required processing speed, by handling the signal through K parallel paths. It is commonly admitted that the price to pay in this case will be an increase of the hardware complexity by a factor of K.

In this context, we considered the option of replacing the GaAs hardware by standard CMOS, with a "parallelization factor" K between 4 and 8. (K is also known as "Time Multiplex Factor".)

Before discussing the possible motivations that could lead to retain this option, let us describe its implementation.

4.1 The high-speed frond-end

The front-end receives the high-speed data (nominally 400 Msamples/sec) and produces a parallel flow K times slower, carrying the same information.

Figure 4 shows the straightforward architecture of this circuit, where each square box represents a couple of D-type flip-flops handling 2 bits of data. It also produces the "slow clock" S_{ck} by dividing by K the frequency of the

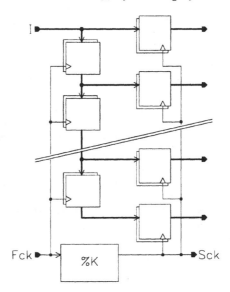

Figure 4 Parallel solution front-end

original (fast) sampling clock F_{ck}.

This part has to be made of high-speed technology, like ECL or GaAs. It represents a very small amount of logic, compared with the complete correlator.

For this reason, standard ECL parts can be proposed in spite of their power consumption, in the other hand an ASIC is not justified for such a simple piece of logic.

4.2   The parallel correlator core

The architecture of the parallel core if derived from the reference architecture, which is the single folded-line system of figure 3, with the goal of producing exactly the same results.

The single folded delay line is replaced by K parallel folded delay lines, clocked by the "slow" clock S_{ck}.

These delay lines operate synchronously, but carry images of the original signal sampled at instants separated by the period of the fast clock F_{ck}.

Each channel has to contain K multipliers, that will work in parallel to perform the same amount of computation as a single mutiplier running K times faster did in the reference architecture. Each of these K multipliers has to process a pair of samples with the same time difference, characteristic of a given channel, and these pairs are chosen in such a manner that the products computed simultaneously are the same as those which are processed in sequence by the single multiplier of the reference architecture.

The K products are added together to feed the integrator.

Figure 5 shows an example of interconnections between a channel and a slice of the K folded delay lines, in the case K = 4.

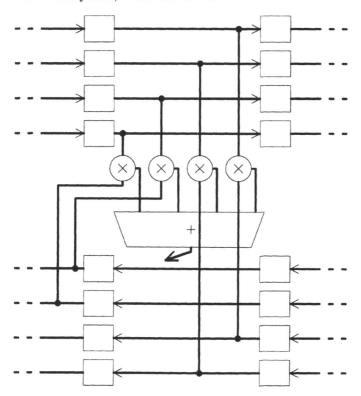

Figure 5 One channel inside the parrallel core (K=4)

Attention is called by the difficulty of representing such a network on a schematic diagram. As soon as the number of represented channels exceeds 3, the diagram becomes unreadable (let us recall that we propose a minimum of 64 channels per chip). The fact is that the structure is still very regular, but this regularity would be manageable only on a N-dimensions schematic diagram!

The solution is to abandon the schematic diagram, and to take advantage of the modern methods based on HDLs (Hardware Description Languages).

We proposed to generate the parallel core by "procedural instanciation", which means that the structure is described by a construction algorithm rather than by a 2-dimension graphical view.

A benefit of this method is the "genericity", allowing to change the values of K or the number of channels without having to rework the circuit by hand.

4.3   Full custom against standard cells in strategy # 2

The main motivation for full-custom design in the case of the GaAs chip of strategy # 1 was the regularity of the structure, easily translatable into regular masks layout. In the case of strategy # 2, the network structure is very difficult to implement by hand in form of a 2D geometry. Moreover, a full custom implementation of the parallel core would lack flexibility, locking the project to a specific value of K.

For these reasons, this time the full-custom approach loses.

The generic procedural generation fits better with the automatic "place-and-route" software that works with a standard-cells library.

Our preliminary experiments showed that the state-of-the-art place-and-route programs, based on a non-deterministic optimization algorithm (simulated annealing) have no difficulty to produce very compact layouts.

5   COMPARISON BETWEEN STRATEGIES

The study of strategy #2 (all-silicon) was motivated mainly by the temporary difficulty to obtain GaAs prototypes and some pessimistic predictions about the future of this technology. In fact the all-silicon approach offers the advantage of making the project much more flexible in the terms of relations with manufacturers.

This flexibility is enhanced by the use of procedural generation and automated layout, allowing to change the foundry and to adapt the K factor to the available performances without a lot of rework.

Another benefit of the all-silicon approach is a simplification of the instrument architecture, because the boundary between the "fast" part (GaAs) and the "slow" part (CMOS) of the correlator dissapears (see figure 2). In strategy #2, these two parts are merged into one chip, and the complex multiplexing scheme used between the fast chip and the slow chip no longer exists. One may object that a new boundary appears between the front-end (ECL) and the correlator core, but this boundary is much simpler to manage than the previous one.

The all-silicon approach has also three major drawbacks: the difficulty of predicting the timing performances, the difficulty to predict the power consumption, and the radiation sensitivity of CMOS. The first difficulty is related with the random place-and-route, the second one is a characteristic of CMOS.

Let us recall that GaAs-MESFET is appreciated for space application, because its power consumption is easy to predict, and it does not produce latch-up in presence of radiations.

6 STATE OF THE PROJECT

Started earlier, the strategy #1 is more advanced. A first prototype of the chip set has been designed, manufactured and tested [7], having shown the feasibility of the instrument but leaving room for many improvements, and an improved version of the GaAs chip is in the manufacturing process.

Fig. 6 is a photograph of the first GaAs chip (area 19.5 mm2) in its special high-frequency ceramic package.

In the other hand, the second strategy has not yet been implemented in silicon. A generic procedural generation program has been developed, producing structural netlists in Verilog format, that can be imported in the CAD framework used for placing and routing the standard-cells. A high effort has been made for verifying by simulation the generated circuits, by means of automatic comparison with the outputs of the reference circuit.
But the speed performances evaluation and the power consumption estimation of the parallel correlator are still to be done.

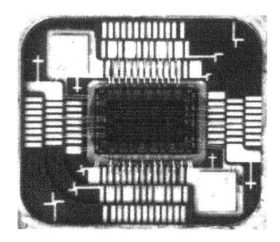

Figure 6   The first GaAs correlator chip

7   CONCLUSION

Two strategies are actually developed for the space spectrometer. The strategy #1 is presently prefered, mainly because it is the most advanced in terms of schedule.

But such a long term project cannot rely only on a solution depending on a unique manufacturer, a second source is mandatory; and this second source is based on strategy #2

Acknowledgements

The authors wish to acknowledge the outstanding contribution of Jorge Sarmiento from the University of Los Andes (Bogota), who inaugurated the procedural generation.

8   REFERENCES

[1] W. J. Wilson, "Radiometer Spectrometers for Space", in "Coherent Detection at Millimeter Wavelength and Applications", pp 163-179, Nova

Science Publishers, New York, 1991.

[2] S. Weinreb, "Analog-Filter Digital-Correlator Hybrid Spectrometer", IEEE Transactions on Instruments and Measurement, IM-34, No4, pp 670-675, December 1985.

[3] B. F. C. Cooper, "Correlators with Two-bit quantization", Australian Journal of Physics, Vol 23, pp. 521-527, 1970.

[4] J.F. Lopez, V. de Armas, J.A. Montiel, "VLSI GaAs Experience Using H-GaAs II Technology", proc. 4th Eurochip Workshop, Toledo, Oct. 1993.

[5] K. Eshraghian, R. Sarmiento, P.P. Carballo, A. Nunez, "Speed-Area-Power Optimization for DCFL and SDCFL Class of Logic Using Ring Notation", Microprocessing and Microprogramming, Vol 32, pp 75-82, 1991.

[6] R. Sarmiento, P.P. Carballo, A. Nunez, "High Speed Primitives for Hardware Accelerators for DSP in GaAs Technology", IEE proceedings-G, Vol. 139, No2, April 1992.

[7] Noullet J.L., Giard M., Crete E., Lagrange D., Mayvial J.Y., Torres M., Ferreira A., "Digital GaAs MESFET Chip for a Radio-Astronomy Spectrometer", European Design and Test Conference EDTC'95, Paris, March 1995. (This paper obtained the "Best Academic Asic Award" at EDTC'95)

DESIGN AND REALIZATION OF A SYNCHRONOUS COOPERATIVE SHARED ELECTRONIC BOARD

V. Baudin, M. Diaz, P. Owezarski, T. Villemur

LAAS CNRS
7, Avenue du Colonel Roche
31077 TOULOUSE CEDEX. FRANCE
+IUT B de Toulouse II
1 place G. Brassens, BP 73
31703 BLAGNAC CEDEX. FRANCE
e-mail: {baudın, diaz, owe, villemur}@laas.fr

ABSTRACT

This paper first presents a generic service to manage cooperative groups of agents and to manage dynamic formation of subgroups of agents inside cooperative groups. The model used to represent relations between groups of agents is based on graphs. The structure of the cooperative groups changes dynamically in time according to the entries and exits of the agents. The cooperative model has then been applied inside teleteaching groups. To provide remote interactions between a student and a teacher, a shared electronic board has been developed: it gives remote views of applications shared through the board and offers remote control of these applications. Then, a future integration of the shared electronic board is proposed inside a whole cooperative class of students.

1 INTRODUCTION

The advanced technologies based on computers video processing are more and more applied inside the Computer Aided Learning domain. The current commercial products are based on multimedia CD-ROM to handle locally stored pictures. Such CD-ROM based technologies do unfortunately not support the interactions and all the data exchanges that can happen inside a whole classroom: they have been first developed either to learn alone in front of a computer or inside a classical classroom with the help of a teacher.

To realize distributed or virtual classrooms in which the students and the teacher are not geographically located in the same place, several tools, supported by high speed multicast communication networks, are required to ensure the necessary interactions between the members of the group: a visioconferencing tool gives the view of all the present members. Through an audio channel, they can talk as if they were located in the same place. An electronic board is used to share information or documents between the members: for instance, the board can broadcast the teacher's course support to the students. Such a tool can be also used by the teacher to supervise the students' exercises or to give remote control to the shared applications.

The aim of this paper is to present the shared electronic board developed and to describe its integration within the cooperative group model retained to represent teleteaching situations. This article is composed of the following main sections: after reporting related work in Section 2, the model used to represent cooperative teleteaching groups with the membership service that manages the group dynamicity are presented in Section 3. The functionalities, the design and the future developments of the shared electronic board are described in Section 4. Finally, some conclusions and perspectives are given at the end of the paper.

2   RELATED WORK

Group models have been widely studied inside the Computer Supported Cooperative Work area, a new multi- disciplinary research domain that studies work of group of users [11]. Most of the time, models are used to describe the mutual interactions of the cooperative entities or to define their behaviours and possibilities. Some group structures are based on Petri Nets [1], objects [2], actors [6]... But few structured groups have been used and defined inside synchronous cooperations that require the co- presence of all members of the group at the same time. Indeed, only synchronous groups can handle live video exchanged between the members. Visioconferences [ISHII91], [12], teleteaching applications [13], [15] are based on synchronous interactions between the supported group members. To work in common on a same application, sharing tools have been introduced. The first ones (Conference Toolkit [3], Shared X [7]) are built on top of the XWindow protocol. Multi-user applications can be developed on top of the Conference Toolkit environment. A prototype has been defined for sharing any X based application. The X protocol does not handle multimedia data and live video: as a consequence, only graphical applications can be shared through such systems. The MERMAID system [12] contains a support to share applications. Each user owns a copy of the shared application and the exchanged events are globally ordered by the system. At any moment, only a user can send events to an application: this floor control is negotiated between the users. Drawings, texts and pictures can be shared, but not videos. The TeamWorkstation system [8] handles video to share documents inside a teleteaching context. A student makes an exercise inside an area filmed by a camera. A teacher writes on another remote area that is filmed too. The pictures are exchanged through the network. Before being displayed, their contents is analysed an mixed as if they were slides. The two drawings are then superimposed. There is however no remote interaction allowed by the system: the teacher can not directly help the student and can not modify his work. The shared board presented in this paper is based on the exchange of videos that contain the views of the shared applications with the possibility of remotely controlling them.

3   GROUP MODEL FOR TELETEACHING

The model used was proposed by Diaz [5]. Each agent has a set of data for which it is the sole owner; no other agent can modify the data. When an agent modifies its data, it sends the new value to the other agents that need this data to realize the cooperative work.

3.1   Dynamic cooperative groups

A cooperative group is composed of agents, organised to define the relations between the members of the group. The structure of the cooperation is represented by a directed graph, the cooperation graph, as shown in Figure 1. Vertices represent the agents, edges represent the relations between them. Thus, an edge from agent "T" to "S" means that "T" can read some or all of the values owned by "S", as shown in Figure 2. The approach retained for cooperation is those of information sharing between agents [10]. Agent "S" cooperates with agent "T" if it gives or shares some of its information with "T".

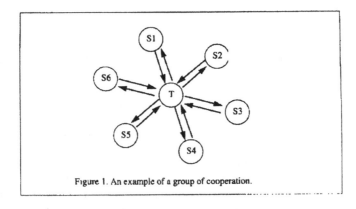

Figure 1. An example of a group of cooperation.

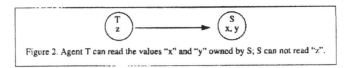

Figure 2. Agent T can read the values "x" and "y" owned by S; S can not read "z".

The cooperation graph defines a structural type, the conceptual structure of the cooperation. This structure can be described in terms of a compound activity, composed of the types of the cooperating entities, the types of information they are allowed to read and the types of information they allow other agents to read, together with a relation between the involved cooperating entities. Let us now consider how these activities are initiated. The trivial solution corresponds with the case where cooperation is considered to begin when all entities are ready to start. Of course, this view is too restrictive as cooperative work can be performed as soon as an adequate and sound set of participants come into existence. This means that at a given instant it is not necessary that all agents are present to start or conduct the cooperative work. As a consequence, the conceptual model is extended to define which agents must cooperate to execute the cooperative work.

The application associated to the cooperative group has to define the subsets of agents which have a meaning for the implementation of the cooperative work. In fact, if a pictorial representation is considered, these allowed subsets of agents form subgraphs of the cooperation graphs. Among all subgraphs of the graph of

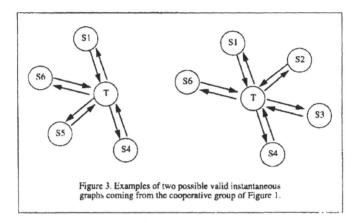

Figure 3. Examples of two possible valid instantaneous graphs coming from the cooperative group of Figure 1.

cooperation, the application chooses those that are semantically possible. The subgraphs retained by the application are called valid instantaneous graphs. Figure 3 shows an example of two valid instantaneous graphs that could come from the cooperation graph of Figure 1.

3.2 Cooperation service and protocol description

The proposed service manages the dynamicity of the cooperative groups defined [14] and aims to pass from a valid configuration where agents are cooperating to another one, in considering the requests of entry in cooperation and the requests to leave cooperation coming from the agents. Let us consider a set of agents which are cooperating and which constitute a valid instantaneous graph. Inside the cooperative group, other agents may want to participate to the cooperative work and join those which are already working. Also, agents which are cooperating, may want to leave the work and stop their participation in the cooperation. Considering the requests to enter and the requests to leave the cooperative work, the service tries to form a new valid instantaneous graph.

Several cases are possible to take the decision of changing the cooperation configuration. The first one is to change each time when possible. The service which manages the cooperation realizes the modification as soon as it is possible. The second one, which has been selected here, considers the decision modification as a cooperative decision. As a consequence, the opportunity of changing cooperation is proposed to the actual set of cooperating agents. The cooperating agents vote to accept or to deny the configuration change.

The service for changing the cooperation structure requires four different phases to manage the interferences between data exchanges and the cooperation structure change. When a cooperation change has been decided, the new cooperating agents must exchange their own initial contexts in following the cooperation structure. The second phase is those of the realization of the cooperative work where agent exchange data when they change their values. The third one appears before terminating a cooperation, i.e. before changing a cooperation structure: the data manipulated by the agents must reach a coherent state. The last one corresponds to the restructuring of the cooperation.

The proposed service has been formally described (Figure 4) using the formal description technique Estelle [9], [4]. An Estelle specification consists of a set of modules, each's behaviour defined by a finite state automaton. To communicate and exchange information between the modules, bidirectional first-in- first-out queues are used. The formal description architecture is based on the Open System Interconnection layer model. Users, which represent the cooperative agents, are connected to modules which provide the service to participate in a cooperative group. The execution of these modules is the protocol which provides the service. The cooperation manager module supervises the evolution in time of the cooperative group. All the protocol modules are connected to another module which provides a multicast service. These specifications have been used to simulate and test our service.

The Estelle code of the protocol modules and of the manager has been reused for an implementation on top of a UNIX platform [14]. Each Estelle module instance has been included inside a UNIX process that have been distributed across an Ethernet network. The UNIX distributed processes implement a generic cooperative group membership service.

3.3   Modelling teleteaching groups

The generic previous cooperative model has been applied to model various teleteaching configurations. Inside such groups, the teacher has a central position (Figure 1). The students hear the teacher's speech and they see his documents required for the course (slides, drawings, notes...). The cooperative

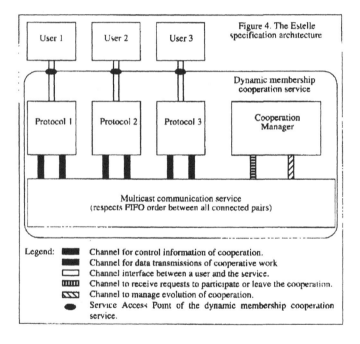

Figure 4. The Estelle specification architecture

graph for a course forms a star whose center is the teacher agent, edges being directed into the central position of the star.

For global discussions between the whole class, each cooperative agent must see the informations of all the members of the group. Such situations require fully connected cooperative graphs. In the case of practical work, the cooperative teleteaching group of Figure 1 can be retained. Students have to make exercises in their own environments. They can not communicate between them to avoid copying. The teacher must have access to each student's context to supervise, help or respond to their questions. The teacher's help can consist in a simple teacher/student dialog but, can be composed of a direct remote modification of the student's context made by the teacher and supported by tools as a shared electronic board. Inside a course, to describe more interactive situations, the teacher can take into account the students' feedback (concentration of the students or questions coming from them). In this case, he must know and see the own informations of the students. This cooperative group is depicted by Figure 1 too. When students ask questions, they can influence the teacher's context with his agreement in remotely pointing or clicking to the subject of their comments. For instance, if a student does not understand a particular point of a drawing made by the teacher, he remotely points the unclear part of the drawing while he asks his question. Such a functionality is taken into account inside shared electronic board applications.

All the valid configurations of a course or a practical work contain the teacher. Some variants can appear inside the choice of the valid configurations: a graph is valid when at least the teacher and one student are present, or when at least the teacher and a fixed number of students are here... The teacher alone accepts or denies the change of a current valid configuration. The interactive course configuration has been used to describe the functionalities of the proposed synchronous cooperative shared electronic board.

4   SHARED ELECTRONIC BOARD

4.1   Presentation

Work presented in this section describes a generic tool for remotely sharing various multimedia applications. A majority of multimedia computer aided learning applications include digitized video sequences to detail or to illustrate several important parts of a course.

As an illustrative example, training pilots and maintenance staffs are made inside aircraft industries in using a Computer Aided Learning System (CALS) that contains video sequences. When a trainee presses a button inside the aircraft cockpit drawing, the system displays other graphics to explain the functioning of an electrical, electronic or hydraulic circuit, this functioning being strengthened by the display of a live video sequence. To learn all the complex procedures required to maintain or pilot an aircraft, pilots and maintenance staffs use CALS and are supervised by an instructor. The trainees and the instructor directly dialog because they are located in the same classroom. Now, let's suppose that all the members of the class are not located in the same place and are geographically distributed: to communicate and to exchange information between them, the group members need workstations connected to a network. Moreover, the distributed information exchanges are only possible if tools that ensure remote dialogs (as visioconferences) and remote interactions or control are available: a distributed electronic board is then useful to share the access of applications owned by the instructor or the trainees. The shared electronic board can be directly applied to the sharing of a CALS that runs on the instructor's workstation. To emphasize the functionalities of the electronic board (Figure 5), the following situation has been chosen:

- a) The instructor owns the CALS that runs locally on its workstation. Through the electronic board, it broadcasts the current picture (that corresponds to the current state of the application). Then, all the trainees show the picture of the instructor's CALS.

- b) During the course, a student asks for a question. When the instructor agrees, a remote pointer controlled by the requested trainer appears inside the instructor's board. This pointer helps the trainee to explain and to show the unclear current displayed parts of the CALS application. The instructor has given to the trainee the pointing right on its electronic board.

- c) For more complete discussion or explanations, the instructor can give to the requesting trainee the remote control right of all the applications viewed inside the electronic board: in our example, the local instructor's CALS is then remotely controlled by the trainee. This can be useful if the question concerns precise points that require the CALS running. While the trainee asks his question, it controls the CALS.

Each member of the group owns a board inside its local screen. A local board is composed of a window that contains views of the shared applications. The

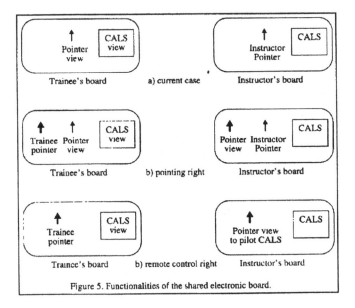

Figure 5. Functionalities of the shared electronic board.

board of the instructor and those of the trainees dialog between them to synchronize their local views of the applications and to ensure the remote interactions between a trainee's board and the instructor's board.

To share and show a view of an application, the instructor slides inside the electronic board window the window application. Any application that the instructor owns can be shared through the electronic board. A trainee sees a view of the shared application inside its local board.

To obtain the pointing right, a trainee puts its pointer inside its local board. Then, once the right has been given by the instructor, two independent pointers appear on the instructor's and trainees' boards, the first one belonging to the requested trainee, the second one belonging to the instructor. Each member controls the position of its own pointer. The applications shared inside the board remain to the control of the instructor. In the remote control case, only one pointer remains inside the boards. The instructor gives the remote control right to one of its shared applications to a trainee. The authorized trainee remotely controls with this unique pointer the chosen application. When the trainee has finished his speech, the instructor takes the application control again. At a given instant, at most one person controls the shared applications: either the instructor or an authorized trainee.

4.2   Realization

The first implementation developed has been simplified and supports only interactions between the instructor and a single trainee.

Platform description

The implementation platform, depicted on Figure 6, is composed of two Sun Sparcstations equipped with a Sun audio card and a Parallax video card. The audio card records and plays back sounds in the PCM format. Pictures can be digitized from two different video inputs, displayed, compressed and uncompressed in real time with the JPEG compression algorithm by using the video card. A 10 Mbps Ethernet local area networks is available inside this platform which provides sufficient throughput for this kind of application.

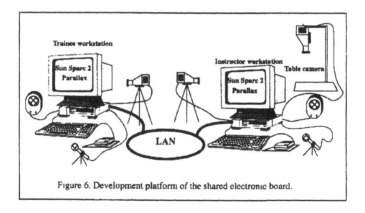

Figure 6. Development platform of the shared electronic board.

Electronic board characteristics

The first video input of the instructor's workstation is used by his shared electronic board and pictures coming from this source are displayed in the background of the board, the shared applications being displayed in the foreground. The background pictures, as the shared applications, are displayed inside the trainee's board. This first video input is connected to a table camera used to grab papers, slides, or any other object displayed to the remote boards. With this additional functionality, the shared electronic board is well suited to display classical conferences supports to the trainees. The second video input is not used by the electronic board application, and can serve for other video applications as a visioconference (with the connection of a camera). The information exchanges between two boards are mainly based on video transmissions: all the contain of the instructor's board (background video and shared application window) is grabbed, digitized and sent to the remote trainee's board. Only a view of the running shared applications is sent to the trainee. This choice has been made for several reasons:

- There is only a local copy of the running applications (that is easier to maintain and to change than several distributed copies).

- Multimedia applications as Computer Aided Learning Systems require a lot of disk space for storing the multimedia data. Distributing a copy of the application to each trainee could not be conceivable. So, the application is only stored inside the instructor's computer.

- The synchronisation problems between the application views sent are easier to manage.

The main problem of sending videos is the rate required by the application. This problem is nevertheless limited with the use of compression technics that often efficiently decrease the amount of data to be transmitted. In the case of very low bandwidth networks, the hypothesis of distributing the shared applications code has been considered and only the events for the application running are exchanged. The copies of the shared application must evolve synchronously and must all receive the same events in the same order. As a consequence, these environments as SharedX [7] require complex algorithms to rearrange all the receiving events before being executed by the copies of the applications.

The communications between the remote electronic boards are based on top of the UDP/IP protocol that ensure the rate needed for video transmissions. The application boards for the instructor and for the trainee have been developed using the C language, and they use the Sun multithreading mechanism with lightweight processes to solve the synchronisation needs of the concurrent threads. Indeed, inside each board, several data flows have to be synchronized (for instance the current position of the pointers and the display of pictures) before being presented. Each board is composed of 1500 lines of C code and

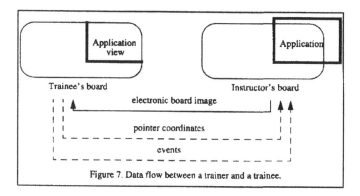

Figure 7. Data flow between a trainer and a trainee.

contains 2 threads for each mode, the first one for receiving the data from the network, the other one for displaying synchronously the data.

Figure 7 gives the data flow between the instructor's board and the trainee's board. In the three functioning mode, the image of the electronic board is always sent from the instructor to the trainee. When the pointing mode is activated, the trainee's pointer coordinates are sent to the instructor's board. During the remote control mode activation, the trainee's pointer coordinates and all the local events created inside the trainee's application using the views of the application are sent to the instructor's board.
Remark: if the window of a shared application is not completely included inside the instructor's board, only the included part is transmitted inside the video and received by the trainee's board.

4.3   Future work: a cooperative shared electronic board

The next step of our work is to extend the electronic board to a whole class of trainees. This requires to take into account the cooperative structure of the group and its possible dynamicity. As a consequence, the general cooperative service and the extended shared board have to be integrated. The proposed general architecture is depicted in Figure 8.

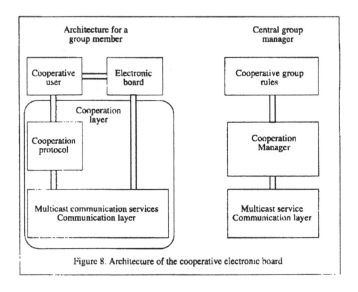

Figure 8. Architecture of the cooperative electronic board

The integration of the shared electronic board and the cooperation membership is done at the application level defined on top of the cooperation layer. The cooperative users are connected to the cooperative service. They request to enter or to quit the cooperative group and they inform the electronic board module of the group structure modifications. The cooperation protocol and the cooperation manager define the cooperative service and they together manage all the modifications of the cooperation. The electronic board part ensure all the functionalities of the cooperative board. The new electronic board modules are now connected to a multicast service to communicate the board data to all the cooperative members. Moreover, the new electronic board of the instructor must take into account several requests that could come from the trainees.

5   CONCLUSION

This article has first presented a general model to represent cooperative services, then has described a generic service that manages the dynamicity of these cooperative groups. A shared electronic board application useful in a context of a teleteaching environment has been presented followed by the proposition of a future integration of the membership service and the board application.

The cooperative membership service has then been implanted using a centralized manager. To improve the robustness of the protocol, current work aim to distribute the group management responsibility in considering the cooperation manager as a special token that circulates among the cooperative agents.

The point to point realization of the electronic board is now asymmetric: the functionalities of the instructor are different of those of the trainees. A direct extension could be to consider a symmetric board: the trainee can access to the instructor's context and, if necessary, the instructor could access to a trainee's context in the case of practical works. In the same way, inside the electronic board, the access to the diverse contexts of the trainees allow the instructor to supervise the whole group and to help some of them if necessary.

At end, the electronic board has to be integrated inside a complete teleteaching environment that contains a visioconference to support the informal dialogs between the members of the group, the subjects of the discussion being shared using the electronic board.

6 REFERENCES

[1] A. Agostini, G. de Michelis and K. Petruni. Keeping Workflow Models as Simple as Possible. Workshop on Computer Supported Cooperative Work, Petri Nets and related formalisms, Zaragoza, pages 11- 29, June 1994.

[2] S. Benford and J. Palme. A Standard for OSI Group Communication. Computer Networks and ISDN systems, 25(1993):933-946. 1993.

[3] A. Bonfiglio, G. Malatesta et F. Tisato. Conference Toolkit: A Framework for Real-Time Conferencing. In Studies in Computer Supported Cooperative Work: Theory, Practice and Design, Eds. J. M. Bowers and S. D. Benford, Elsevier, 1991.

[4] S. Budkowski and P. Dembinski. An Introduction to Estelle: A specification Language for Distributed Systems. Computer Networks and ISDN Systems, 14(1987):3-23. 1987.

[5] M. Diaz. A logical model of cooperation. Proceedings of the IEEE. Third Workshop on Future Trends of Distributed Computing Systems, pages 64-70. April 1992.

[6] C. Hewitt and J. Inman. DAI Betwixt and Between: From "Intelligent Agents" to Open System Science. IEEE Transactions on Systems, Man, and Cybernetics, 21(6):1409-1419. November/December 1991.

[7] HPSharedX. Hewlett Packard. July 1991.

[8] H. Ishii et N. Miyake. Toward an Open Shared Workspace: Computer and Video Fusion Approach of Teamworkstation. Communications of the ACM, 34(12):37-50, December 1991.

[9] ISO/IEC ISO 9074: 1989 (E). Information processing systems. Open System Interconnection. Estelle: A formal description technique based on an extended state transition model.

[10] S. M. Kaplan and A. M. Caroll. Supporting Collaborative Processes with Conversation Builder. Computer Communications, 15(8):489-501, October 1992.

[11] A. Karsenty. Le collecticiel : de l'interaction homme-machine à la communication homme- machine-homme. Technique et science informatiques, 13(1): 105-127, 1994.

[12] T. Ohmori, K. Maeno, S. Sakata, H. Fukuora et K. Watabe. Distributed Cooperative Control for Sharing Applications Based on Multiparty and Multimedia Desktop Conferencing System: MERMAID. Proceedings of the IEEE: 12th International Conference on Distributed Computing Systems, pages 538-546, Juin 1992.

[13] P. Owezarski, V. Baudin, M. Diaz and J.F. Schmidt. Multimedia Teleteaching: Introduction of Synchronization and Cooperation Mechanisms in distance learning. Proceedings of the World Conference on Educational Multimedia and Hypermedia, pages 517-522, june 1995.

[14] T. Villemur, M. Diaz and F. Vernadat. Validated Design of Dynamic Membership Services and Protocols for Cooperative Groups. Annales des Télécommunications, 50(11-12):859-873, November/December 1995.

[15] J.M. Wilson, D.N. Mosher. Interactive Multimedia Distance Learning (IMDL): The prototype of the Virtual Classroom. Proceedings of the World Conference on Educational Multimedia and Hypermedia, pages 563-570, june 1994.

IMPORTANCE OF SPECIFICATION MEANS TO DESIGN INTEGRATED MODULAR AVIONICS SYSTEMS

G. Motet

GERII / LESIA, DGEI / INSA,
Complexe Scientifique de Rangueil, 31077 Toulouse cedex, France
Phone: +33 5 61 55 98 18, Fax: +33 5 61 55 98 00
e-mail: motet@dge.insa-tlse.fr

ABSTRACT

Economical constraints such as maintenance cost reduction required the introduction of a new architecture to design systems which will be embedded in the future commercial aircrafts. This architecture called "Integrated Modular Avionics" leads to a new approach of system design. In particular this architecture allows multiple applications to be integrated on one framework. Moreover, as studied systems concern the avionics domain, dependability must be considered as a major property of these systems. During the European BRITE-EURAM project "IMAGES 2000", the main European aircraft manufacturers and suppliers and some research laboratories worked together to study the means and the process which must be used to design dependable Integrated Modular Avionics systems. In this paper, we analyse the characteristics of the Integrated Modular Avionics systems to highlight that the obtaining of dependable systems will require a special effort on specification step from the part of the engineers involved in the design of the systems embedded in the future planes.

1 INTEGRATED MODULAR AVIONICS SYSTEMS

In traditional architectures, the avionics functions are embedded in Line Replaceable Units (LRUs). Each LRU is an item of equipment dedicated to only one avionics function. It is designed completely (i.e., in particular, it needs its hardware resources) and is completely removed in case of failure. With the Integrated Modular Avionics (IMA) concept, a framework, specific to the avionics requirements, provides all the resources (for processing, I/O, power supply, etc.) needed by the avionics applications. The second originality of the IMA concept concerns the implementation of the framework. The cabinets composing the framework are divided into several types of modules: power modules, core modules, I/O modules and a gateway module. The cabinets communicate by using a multi-transmiters network (aircraft bus). The ARINC 651 standard highlights these two aspects in the definition of the IMA concept:

- firstly, it is modular: within cabinets, standardised modules provide all the required resources and communicate through a standardised backplane bus. As these modules are the basic components for maintenance, they are named Line Replaceable Modules (LRMs);

- secondly, avionics functions are integrated: this means that each IMA platform receives several numbers of avionics functions. The resources are thus shared.

The Integrated Modular Avionics concept will allow:

- multiple avionics applications to be produced by different suppliers and integrated on one platform;

- the modules to be designed independently to the functions.

There are economical interests:

- for the airliners: mainly the reduction of the maintenance costs because there are no one different hardware by function but the functions share standardised IMA platform including standardised modules;

- for aircraft manufacturers and equipment suppliers because it is not necessary to design again the resources which are necessary to execute the functions.

2 IMPORTANCE OF SPECIFICATION STEP IN IMA CONTEXT

2.1 Introduction

In one sub-task of the "Images 2000" project, we studied at what moment in the software system life cycle, faults are introduced. Moreover we tried to evaluate the rate of the faults introduced at each step. This study was not specific to IMA systems but concerns any software/hardware application. The goal of this study was to signal the development steps on which efforts must be focused to avoid the presence of faults in any software/hardware systems.

This study showed that numerous faults included in software do not come from problems associated with design and programming phases but are due to a bad expression of client's requirements as specifications. The rate of 30% of the faults is provided by [7]. This rate value is also given by other authors [1]. These studies concern general software/hardware systems, that is they are not specific to avionics area. In reality this value is higher. Indeed, at each design step, the designer must express the specifications of entities (components, functions, data, etc.) which will be designed in the next step. The designer is therefore his or her own client [14]. Thus, he or she introduces additional faults during design step which are however relative to specification activity. For instance, [6] and [3] quoted that 25% of the faults occurring during the design phase correspond to problems associated with the interfaces of components used during this phase. Other pieces of information reinforce this opinion: [17] specifies that 30% of the faults come from the fact that the limit values of data or the states not frequently reached are badly taking into account by the designers. This value is higher for systems for which interactions with hardware or software components are numerous. For instance [15] presents a control software in which 56% of the faults are coming from problems of interfacing with software components (36%) or hardware components (20%). So, certain characteristics of the developed systems may lead to the increasing of the number of the faults due to erroneous specifications. Importance of component interactions is one of these characteristics which exists in particular in IMA systems.

In the following sub-sections we will present five characteristics of the IMA systems (multiple partners, multiple domains, multiple interactions, complexity of the behaviours, maintenance) showing both the requirements and the difficulties to obtain correct specifications. These characteristics therefore increase the risk of presence of faults in specification documents. So this paper concludes that important effort will be required on specification elaboration to obtain dependable IMA systems.

2.2   Multiple partners

The IMA concept requires multiple plane manufacturers and suppliers to work together in order to design, produce and maintain hardware and software systems embedded in planes. Such a cooperation previously existed between European manufacturers for instance for the Airbus aircrafts. However it was not so strong because each of them had to develop separate and (relatively) independent parts of the control system embedded in the planes. Now, the IMA concept requires the sharing of common resources provided by the framework, then it implies a strong integration of the developed elements and therefore a strong inter-dependency between the partners. It is expressed in [10] as follows: "Equipment suppliers are no longer delivering pieces of hardware and software which operate as a complete unit. Mostly they supply software units which need to be integrated with other units, probably by a third part, before being complete". Such a sharing exists at applications levels (sharing of the platform) but also at module levels (modules must cooperate) and at the level of the components of each module.

For instance firstly [11] defines the resources shared by a core module (processor, memory, backplane bus, operating system, etc.). Consequently, the expression of the precise specifications of the IMA components or modules is absolutely necessary because persons of various companies will have to share IMA resources to develop their parts of the global system elements.

Secondly, multiple partners have also to work together due to the modularity of the framework structure. Indeed the framework is split into modules cooperating to provide the global service offered by the framework.

The existence of multiple partners is also the cause of the first difficulty to obtain specifications. On one hand each firm has one proper way and means to express its specifications; this concerns the style of the specification. On the other hand numerous pieces of information are considered as implicitly known (they belong to the culture, i.e. the basic knowledge, of the company memberships); this concerns the contents of the specifications.

To conclude this first aspect: on one hand, multiple partners have to work together to design applications sharing common resources, so they require common specifications of the framework; on the other hand, multiple partners have to work together to design components cooperating to provide the services of the IMA framework, therefore they also require common specifications, now for these components.

2.3 Multiple domains

Concerning the IMA framework design, an original structure was chosen. The cabinets which compose the framework are split into modules cooperating to provide the global framework services. The originality comes from the fact that these modules are not associated with functional parts but resources. Then, the IMA framework cabinets contains one or several of the following modules: power supply module, core module which provides computation resources, I/O module for external communication needs and gateway module for bus plane communication needs. The second characteristics of IMA systems is thus the presence of several technological domains. Due to the required cooperation of multiple partners, the designers must understand the concepts of these domains, without being a specialist. For instance the following four considered types of modules concern four domains:

- Power Supply module requires knowledge in electricity and uses the associated terminology;

- Core module deals with process management and so handles terms such as "synchronisation", "scheduling", etc.;

- I/O module explanation assumes knowledge on OSI layer model;

- Gateway module treats of one more subject.

So the expression of the specifications of IMA modules is necessary to allow people working in various domains to cooperate. However, a person working on one domain is not able and does not have to know detailed information on the technology used in other domains. However he or she has to possess a complete knowledge on the behaviour of the other elements of an IMA platform, that is on their abstract specifications.

So, the presence of multiple domains at the same time:

- requires the existence of abstract specifications allowing concepts to be handled without knowledge on their technological implementation;

- makes difficult the production of these specifications because it does not authorise assumption of implicit knowledge as it is frequently assumed by the persons working on the same domain.

2.4   Multiple interactions

Another characteristics of IMA architecture is the strong interactions between the elements (modules or components) and also with the applications. All the IMAGES 2000 reports specify numerous relationships between these elements.

In the IMAGES 2000 project, the interactions are at first required by the **definition of the functions** of the IMA elements: evident interactions exist between the I/O module and the core processing modules because the I/O modules aim to interface physical signals and logical events and data processed by other modules. The definition of the Health Monitoring role gives another example of interaction requirements. "The Health Monitor distributes information about the failures of the components to other components (...) The Heath Monitor consists of a distribution and a membership part" [9]. Numerous other interactions between elements may be signalled: intra cabinet communication and inter cabinet communication, communication between the Health Monitor and the Core module for reconfiguration or to communicate information about the state, etc.

The interactions are also due to:

- the choice of a **distributed implementation** instead of a centralised one. An example is the distribution of the error handling: the detection is done in various components (processor, executive, application) but also provokes reactions in other parts (local health monitoring, leader health monitoring). Another example concerns the core module: for instance the synchronisation management may require communication protocols between components;

- the implementation of **fault tolerance mechanisms**. For example, the replication of elements requires the communication of information, for instance to compare the data produced by several versions (N-versions technique). In particular to confine the occurrence of the errors in order to master their effects, the system must be split (concept of segregation). The interactions of the introduced parts must then be specified. Another example is the health monitoring. In order to have the monitoring of errors in one location, the occurred errors must be communicated to the health monitor;

- the implementation of **abstract services**. This reason conducted for example to the partitioning of the Core software and the introduction of the HIS (Hardware Interface System) "which maps the logical requests made by the executive onto the particular physical configuration of the core module" [9]. In the same way APEX interface was introduced which creates a splitting into several elements and then interactions between them (see ARINC 653). Another example is the "logical communication object" [12] used to communicate between the Local Health Monitoring and the Leader Health Monitoring and called "port" in the ARINC 653 document. Let us note that definitions of abstract services allow also specifications which are independent to implementation technology (hardware or software, centralised or distributed) to be defined as previously required.

On the opposite, the previous facilities make difficult the definition of specifications. For instance, the abstract definition of the inter-partition communication hides the means used to implement the real communication. For one way used to communicate, the required time may be defined. However, this way could change dynamically: "for example, two partitions may communicate over the intra-cabinet bus, or even over the inter-cabinet bus, in one configuration, but may be reconfigured to communicate while on the same core module. (...) The end result is that the transmit time of communications might be widely varying in some situations, but the computer must accommodate this" [11]. This last sentence implies that communication time must be specified by an interval [Tmin, Tmax] and not by one value T.

The specification of the interactions between IMA elements composing the platform is specially important and complex because the controls of interactions are not conventional. For instance, there are not hierarchical relations between elements as it exists for sequential systems (sub-program interaction model). Moreover due to the segregation requirements (in particular the faults occurring in an element must have no effect on the behaviour of another one), all the possible interactions must be mastered (no hidden interactions).

Specifications of the available interactions between the IMA platform and the IMA applications are also required. A reference manual defining the IMA platform specifications must be written.

Numerous authors showed that the number of faults generally increases when the number and the complexity of the interactions between components increase. These faults may concern the interface [3] [6] or the behaviour, such as states not frequently reached and therefore badly taking into account [17]. So, in order to obtain a dependable IMA system, the interactions of the modules and between the platform and the applications must be mandatory well specified.

2.5 Complexity of the behaviour

Another characteristics of the IMA system is the complexity of the behaviour of the components. Indeed the numerous interactions induce numerous internal states and thus numerous different cases of behaviour.

To reduce the complexity of a system, the specification of its behaviour is frequently split. For instance to define the Process Scheduler, the project [11] had proposed to split it into several elements: process identification, semaphore management, event management, buffer management, process queue management, time and scheduling management. However each group of services cannot be specified independently in a simple way because correlations between the groups exist. For instance when an event occurs (event management service) to resume a waiting task, the queue management module is called (the task goes from the queue of the waiting processes to the queue of the ready processes). So the complexity of the behaviour of the numerous services of the components cannot be handled, examining each service one by one, independently to others.

Unfortunately, numerous studies showed that the increasing of an application complexity implies the increasing of the number of faults [1], [2], [8], [16], particularly due to erroneous specifications. So complexity is another reason needing to pay a special attention to the specification. In practice, the behaviour of the framework as well as the ones of the modules will be relatively simple to obtain the mastering of the IMA systems.

2.6 Maintenance

Finally the IMA architecture aiming to facilitate the maintenance by changing elements, the specifications of these elements are very important because they define the interoperability of the elements and then their interchangeability. This aspect concerns multiple technological means: mechanics, hardware connection, electrical compatibility, hardware and software communication, operational means (hardware and software operating system), etc.

In particular the definition of interchangeability criteria requires the specification of numerous temporal information. For instance a study [10] dealt with time constraints for temporal segregation. Let us signal that we highlighted in another project the difficulties to specify this kind of temporal pieces of information [13].

Another aspect concerns the error handling. The IMA architecture requires the communication of the detected errors to the health monitoring [12] for a maintenance objective. Thus, the interchangeability criteria includes the specification of the errors which may be transmitted. The definition of an exhaustive list of errors is not easy because some of them depend on the design choices or the technological choices used to implement the component.

To facilitate the maintenance, the specified elements must be generics as possible. Moreover, the "good" level of standardisation must be chosen.

3 CONCLUSION

In this paper we did not describe any solution; we just highlighted problems. However, the given information may be used to provide three pieces of advice.

First and foremost *common means* must be shared by aircraft system designers intervening in the development of IMA systems *to handle specifications*. This is necessary to take the five described characteristics into account. These means concern: expression languages, tools (for checking the completeness, etc.) and methods (to produce specification expressions).

The second advice concerns the common methodology used to specify IMA elements. *The designers of IMA elements must be warned about the great risk of fault introduction in the specifications due to the type of the system to be developed. To highlight this warning, the giving of the values resulting from the references previously quoted must be communicated to these designers.* Thus the engineers will perceive the necessity to use seriously the languages, tools and methods which will be proposed.

The work to be done to write the specifications may be long. This moment is often perceived (for a part) as lost time which will provoke delays in the project schedule. In fact it saves time. Indeed studies show that, if all the faults introduced in the development of a software tool (what ever are their origins) multiply the costs by two, the same studies also signal that the correction cost of a fault due to bad specification is 100 as many expensive than the cost of the studies of the specifications which would allow the detections of the fault to be obtained [4] [5]. So, the use of the techniques proposed to avoid faults in IMA components specifications will cost time and therefore money but less than if they are not used.

The bad perception of the work to be done to obtain the specifications comes from the fact that, at first, the managers often focus on the system disposal and after on its quality. On the opposite, in Japan, the obtaining of a consensus between the client and the designer is very important; then this phase may be long. [18] gives two examples of two big software applications for which the specification expression and the preliminary studies consume 60% (first example) and 70% (second example) of the time necessary to obtain the software tools. He quotes that this long work then allows a very quick design and programming. This short time is due to the fact that the designers were impregnated with the client application domain.

So, the third advice to be included in the common methodology used to specify IMA elements is the following one: *the IMA project managers must be warned about the importance of the requirements specification phase. The use of means to obtain dependable specifications costs money but saves more. To highlight this warning, the giving of the values resulting from the references previously quoted must be communicated to these managers.*

4   REFERENCES

[1] Albin J.-L., Ferreol R., "Collection and Analysis of Software Measurements", (in french), Technique et Science Informatique, vol. 1, no. 4, (1982), pp. 297-313.

[2] Basili V.R., Boehm B.W., Clapp J.A., Gaumer D., Holden M., Salwen A.E., Summers J.K., "Use of Ada for FAA's Advanced Automation System", The Mitre Corporation Technical Report MTR-87W77, (April 1987), pp. 87-120.

[3] Bhandari I.S., Halliday M.J., Traver E., Brown D., Chaar J.K., Chillarege R., "A Case Study of Software Process Improvement During Development", Transactions on Software Engineering, vol. 19, no. 12, IEEE publisher, (1993), pp. 1157-1170.

[4] Boehm B.W., "Verifying and Validating Software Requirements and Design Specifications", IEEE Software, (January 1984), pp. 75-88.

[5] Boehm B.W., "Introduction and Overview", in "Software Risk Management", B. W. Boehm editor, IEEE Computer Society Press, (1989), pp. 1-16.

[6] Chillarege R., Bhandari I.S., Chaar J.K., Halliday M.J., Moebus D.S., Ray B.K., Wong M.Y., "Orthogonal Defect Classification. A Concept for In-Process Measurements", Transactions on Software Engineering, vol. 18, no. 11, IEEE publisher, (1992).

[7] Eckhardt D.E., Caglayan A.K., Knight J.C., Lee L.D., McAllister D.F., Vouk M.A., Kelly J.P., "An Experimental Evaluation of Software Redundancy as a Strategy for Improving Reliability", Transactions on Software Engineering, vol. 17, no. 7, IEEE publisher, (1991), pp. 692-702.

[8] Glass R.L., "Persistent Software Errors", Transactions on Software Engineering, vol. SE-7, no. 2, IEEE publisher, (1981), pp. 162-168.

[9] Sub-Task 2.4, "Requirements on Cabinet Monitoring Aspects in IMA Context", Working Report no BAe_001_WD_2.d of the IMAGES 2000 BRITE EURAM Project, (January 1994).

[10] Sub-Task 5.3, "Guidelines for Segregation Mechanisms Implementation", Official Report no SI_001_OD_5.c of the IMAGES 2000 BRITE-EURAM Project, (January 1995).

[11] Sub-Task 5.4, "Guidelines for Real-Time Implementation", Official Report no SXT_005_OD_5.d of the IMAGES 2000 BRITE-EURAM Project, (February 1995).

[12] Sub-Task 5.6, "Guidelines for Health Monitoring Implementation", Official Report no AS_007_OD_5.f of the IMAGES 2000 BRITE-EURAM Project, (July 1995).

[13] Motet G., Kubek J.-M., "Dependability Problems of Ada Components Available via Information Superhighways", Proceedings of the 13th Conference

on Ada Technology, Valley Forge, Pennsylvania, USA, Rosenberg & Risinger Publisher, (1995), pp. 8-18.

[14] Motet G., Marpinard A., Geffroy J.-C., "Design of Dependable Ada Software", Prentice Hall, (1996).

[15] Nakajo T., Kume H., "A Case History Analysis of Software Error Cause-Effect Relationships", Transactions on Software Engineering, vol. 17, no. 8, IEEE publisher, (1993), pp. 830-838.

[16] Schneidewind N.F., Hoffmann H.M., "An Experiment in Software Error Data Collection and Analysis", Transactions on Software Engineering, vol. SE-5, no. 3, IEEE publisher, (1978).

[17] Sullivan M., Chillarege R., "Software Defects and their Impact on System Avaiblability. A Study of Fiel Failures in Operating Systems", in the proceedings of the FTCS 21, (1991), pp. 2-9.

[18] Tamai T., Itou A., "Requirements and Design Change in Large-Scale Software Development: Analysis from the Viewpoint of Process Backtracking", in proceedings of the 15th International Conference on Software Engineering, IEEE Publisher, (1993), pp. 167-176.

I would like to thank René Meunier who works at Aérospatiale, for his contribution to this paper. This paper stems from a part of INSA participation in the European Brite-Euram project "Images 2000" sponsored by the European Union.

19

INDUSTRIAL COOPERATION: DEFINITION, INTEREST AND DYNAMIC EVOLUTION

M. Filippi

UFR Droit et Sciences Economiques, Universite de Corse
BP 52 20 250 CORTE
e-mail: mfilippi@univ-corse.fr

ABSTRACT

The aim of this communication is to explain cooperation agreement as a specific mode of coordination and to study how this agreement relation is embedded into organisational knowledge and competences. In a first part, we propose a definition of the motivations for cooperation. The second part deals with cooperation. This last guides responses to the unanticipated technological change using "trust".

1   INTRODUCTION

By cooperation we mean any kind of coordination as an objective choice from agents, conscious to create an interelation. The aim of this communication is to explain cooperation agreement as a specific mode of coordination and to study how this agreement relation is embedded into organizational knowledge and competencies. In a first part, we propose a definition of the motivations for cooperation. The second part deals with cooperation. This last guide responses to the unanticipated technological change using "trust".

2 A DEFINITION OF AND MOTIVATIONS FOR COOPERATION

This part is dedicated to the identification and understanding of the specificity of cooperation. Therefore, we have to explain the proper mechanisms of this type of coordination. The analysis of various databases and works dedicated to this subject already let us design, as we will explain it later on that whatever their forms (franchising, joint venture), cooperations keep the identity of each one of the partners, on the contrary of integration. They must be equitable to be stable in time, through the obtention of mutual advantage which does not imply that cooperation involves partners of the same weight. Those three criteria then allow to precisely specify the kind of cooperation and to distinguish them from the other forms of interfirms relations.

The nature of industrial cooperation is linked to acknowledgment of concerted sharing division of labor between the firms. Thus, by essence, this raises the question to render itself dependent faced with an other firm, or even a competitor, which to be justifiable requires a higher profit than the one procured by internalization. Also, to survive, cooperation must rely on equitable partition of the benefits. Commitment and profit evaluation not only requires an efficient control and evaluation system but also the development of one organizational equilibrium and a long term strategy. Consequently, we can wonder what are the original motivations of the choice of cooperation. This question joins the question of dynamic efficiency that must conciliate resources creation and organizational flexibility.

2.1 Nature and Dilemma of cooperation

Throughout cooperation, firm seeks externalizing a part of its activities to reduce the constraint caused by its productions and innovations (due to the presence of set up costs). These last alleviate the flexibility and capability of reactions of the firm faced with environmental changes.

Industrial Cooperation Nature: coordination of dissemblable activities

The question of activities coordination using a strategy other than the one of the market or internalization, joins the question of the nature of industrial cooperation. Promoted by firms, partenariatships are based on the notion of capability of one organization to mobilize its competencies in order to develop productive processes. This notion of capabilities [19] allows to explain why the firm coordinates different activities. The activities diversity used by firms leads them to select those that they practice according to the characteristics of the required capabilities for their implementation. The capabilities cover at the same time the formalized objective knowledge and experience, fruit of learning. Thus RICHARDSON distinguishes the same activities from the one that are complementary [20].

The same activities are those that use the same capabilities in term of knowledge and experiences for their realization inside the firm. Complementary activities can represent various phases of production processes and therefore they require to be efficiently coordinated. "...that activities are complementary when they represent different phases of a process of production and require in some way or another to be co-ordinated" [20]. This activities coordination principle then leads to unite the same and complementary activities into the firm, not similar but complementary activities being coordinated by cooperation. Partnership corresponds to one work division between the firms for which "the root of cooperation agreements seems to be, in fact, that partners agree of obligation degree and devote to assurance degree for the respect of future behavior" [8]. Then, cooperation exists if two or more organizations agree on a ex-ante production strategy. This allows then to coordinate their complementary but dissemblable activities (quantitatively and qualitatively). The aspect of voluntary ex-ante coordination is important in the way that it devotes conscious effort to reach together a collectively fixed goal. But this choice is not riskless.

Dilemma

To cooperate is not an innocent behavior but it is based on a dilemma. The adoption of one cooperation is a strategy that relies on delicate choice: to ally with a rival or not implies to limit future opportunities. The firm must manage activities by specialization, integration or diversification, keeping at the same time, an internal organizational equilibrium, measured throughout flexibility. "In its largest meaning flexibility translates the possibility for a manager to be able to consider at anytime once again his choices in order to maintain the optimality of his decision" [2]. Faced with an organizational choice, where coordination can be carried out using alliance or internalization, this puts forward the question of loss or maintain of future possibilities, in other words the question of choice between static or dynamic flexibility compared to the environment of the firm. Thus, the static flexibility depends on the existence, at a given time, of a set of more or less important number of opportunities.

The aim of the firm is to constitute an optimal range of products. The use
of cooperation strategy, that joins the choice of an external coordination of
activities, can facilitate this goal by cost repartition among partners. FORAY
[8] points out that when coordination depends on interfirms cooperation, it is
then possible to allocate sunk cost amongst several organizations. Cooperation
allows then to release a double constraint:

- to decrease the sunk cost linked to the investments,

- to develop the specificities (human resources) either by adaptation or by
  integration of technology.

But the firm's behavior can have a static position by the answer aspect of the
firm's behavior. However the interest of the choice of cooperative strategies
seems to be more interesting for the implementation of an initiative behavior.
Because environmental firms changes, these last must try to anticipate changes
in order to have a more efficient capability of reaction. Thus, the dynamic
of organization relies on contradiction between the integration necessity and
resource association to give them specific (technology creation condition) and
the necessity to put them on the market (reversibility condition) [8]. This compromise takes as much importance as the environment of the firm is perturbed
(strong uncertainty) under the effects of the technological development. The
choice to prefer an internal coordination to the external one then implies the
question of the dynamic efficiency.

2.2   Dynamic Efficiency question

Cooperation strategy'choice is for the firm a complex problem because it is
linked to the notion of dynamic efficiency in other words the capacity of the
firms to create new resources keeping environmental dynamic flexibility. In this
case, cooperation is a difficult but advantageous solution.

To create new resources

Cooperation choice is often viewed by the firms as a loss of freedom towards
its competitors. This is the raison why from an internal organization's point
of view the capacity of the firm to create and develop new resources is linked
to the notion of competencies as firm'core business that enable it to preserve
its comparative advantage. Then, a first, extension carries out by the strategic

management developed it-self with WILLIAMSON's works on internalization linked to the notion of assets specificity in order to over come its failures concerning the integration of innovation and technical change. RUMELT [22], for example, defines the firm as a capability or unique resources serial, combined in order to accordate itself to the demand changes. But, other authors, in an evolutionist trend using learning phenomena, explain dynamic efficiency using learning processes. From that moment, organization structuration is no longer the cost minimization but the possession by firm of specific competencies and rules. "The organizational firm's essence is linked to the basic competencies that it has in other words a set of technological competencies complementary assets, rules, that identify the firm in a given activity" [10]. Then firm's core business is strategic core of the firm. This is the raison why "a basic competence is a set of technical differentiated competencies, complementary assets, and rules, that altogether form the basis of currential capabilities of a firm in a particular activity.... Typically, such differences have a major underlying dimension that renders the imitation by the other difficult even impossible" [10]. These authors defend the idea that a distinction between complementarity competencies (that may be externalized) and crucial competencies (that firm's core business allows to maintain a internal coherence of the organization). The organizational goal is more, today, to produce specificities than to develop products. But, the increasing uncertainty makes difficult this goal realization as the organizational learning, sources of modifications of knowledge and rules can also be a destabilizing tool for the organization. For COHENDET and LLERENA[2], "this is an implementation of localized learning processes that constitute the most adequate organizational answer for the viability of an organization". If organizational learning is likely to explain the difficulties of the activity coordination for internal organization or alliances, it may be significative of dynamic flexibility in order to answer to the question of efficiency.

To develop a dynamic flexibility

In order to keep its internal coherence, managing the organizational internal coherence compromise, the firm develops cooperation strategies in order to find an equilibrium between internal and external coordination of its activities. But, in this context, it is not longer a static but a dynamic flexibility. This capability to continuously react in the time, to environmental variation will be implemented generated by the external environment or by the capability to improve future choices. In the case, organizational flexibility is dynamic because it develops knowledge accumulation processes. According to FAVEREAU [6], it is this capacity of stimulation and orientation of learning, capability that have or that can have organization in order to maintain or to increase the possibility to react at the change. To stress the importance of environmental uncertainty is the same as to ask why complexity and irreversibility management are objectives of the firm. Complexity and irreversibility are united by the introduction of uncertainty in the analysis. According to AMIT and SCHOMAKER [1], managerial decisions must take into account uncertainty (technological, competitor's behaviors and customer's preferences), relational complexity (competitive interactions) and inter-organizational conflicts. These three conditions leverage decision making of individuals and their important choice. Irreversibility is one of the most important characteristic of decision making in a non probabilisable uncertainty and more generally speaking one aspect of the phenomena linked to the development of innovations. More precisely, environment generates complexity and uncertainty due to three sources of possible non probabilisable uncertainty: innovation, others agent's behavior and environment it-self. Cooperation, then is a form of organization that allows partners to reduce those disruptions. By its organizational mode, more flexible and independent of the major organization, it is an organization form that manages interface with the environment. In other words, it carries out a test (for new products, new technologies and new partners). Cooperation agreements create compatibility area. Thus, they are particulary useful for idiosyncratic resources merger with an important tacit constraint. Depending on propriety of loose coupling, interdepedencies are strong inside the system and these are weak inside intersystems. Consequently, this explains the cooperation characteristics as an organizational form allowing to the firm to develop new organizational forms without bringing disruptions inside them. Cooperation is partners association in a more flexible setting that it is, for example, the one of the firm.

Thus, cooperation manages uncertainty and complexity keeping the own identity of partners and allowing the assimilation innovations (product but also the most organizational), once functioning rules are routinised. It then becomes obvious that cooperation stake is the process learning elaboration and the development of rules. As LE BAS [13] says as the firm is a learning laboratory, cooperation agreement can be considered as an experimentation field.

3   COOPERATION DYNAMIC

Cooperation evolution puts forward two important points: organizational learning and trust. The following hypothesis discussed is: organization learning among partners is an indispensable tool for a good evolution in the time of the cooperation relation. It develops itself among voluntary agreements, that are contratualised or not. It is a support of trust built during partnership, because if learning exists during cooperation, actors win mutual assurance by better mutual knowledge with exchange, using also rules and mutual habits. From that point, we can consider it as an specific asset, in other words as an output of relation. We propose to precise the notion of organizational learning before linking it with the trust one.

3.1 Creation and Development of organizational learning

Learnings depend on a great part on the nature of accumulation knowledge and the way knowledge is developed. "In these conditions, technology is not a public good, but it implies specific knowledge, idiosyncratics, partly approppriable, that are accumulated in the time throughout learning processes, specific too, whose the top managers depend on the proper knowledge of firms and technologies already used". Then, we will precise knowledge characteristics and their influences on organization. "Search activity, and much more again the one of development, leads to an accumulation process by knowledge, learning and how-know that is not completely formalized and come within individual or collective practices" [11]. Usually, we agree to recognize the following technological innovation characteristics: "Technological innovation is process that lasts differently according to industries. Moreover it is strongly localized, tacit, path dependent and it has irreversible process characteristics" [3]. To understand the stake that constitutes the organizational learning creation for the cooperation choice viability supposes to recognize two characteristics of knowledge: tacit and cumulative aspects. Therefore, the tacit aspect of internal rules of a firm can be such as a firm, trying to imitate its concurrent, will have the most important difficulties to do it, if the firm has no access to those rules. From then, the organizational context is fundamental. The boundary between explicit and tacit knowledges is given by the practice and the strategies developed by the actors. So, the organizational change is strongly linked to innovation. If for KANTER, innovation is a process allowing to find new concrete solutions in the firm, MEZIAS and GLYNN [15] define innovation as a discontinuous, non daily and significative organizational change. This is the reason why, the implementation of new organizational structures must rely on confidence, reciprocity, in other words it has to be coordinated by cooperation. The importance of interactions between individuals plays a major part.

The tacit feature of knowledge puts forward the necessity of a stabilized frame in order to favour the repeating and hereby the acquiring of codes to preserve the knowledges but also the necessary components of their diffusion.

According to DOSI, TEECE and WINTER, "what has been learnt during a period relies on what was learnt during past periods" [4]. The accumulation aspect of learning is fundamental at the individual and organizational level to yield past experiences. Therefore, there is learning when the technical change can be considered as a result or a step in the following building process, that is to say a regulating process improving the existing technical solutions at technology structure of unchanged basis [18]. However, the cumulative aspect outlines also the fact that: "the output of researches is not only today a mere new technology but also the submission of knowledges and this output constitutes the basis of new blocks to be used tomorrow" [16]. Learning is irreversible and justifies the choice of cooperative strategies. LE BAS and ZUSCOVITCH [23] define the path of learning depending on each firm, as the combination of internal and external learning and the technological way as the configuration of those paths. From then, the accumulative part of learning is to structure the firm. Coming from its cumulative feature, a localized feature can be given to knowledge. As a matter of fact, we know the importance of tacit and accumulation (and its accumulation mode inside the organization) when building a knowledge. The importance of the organizational structure, of the information system and the way of taking decision render the combining resources mode as a major tool in the development of specific knowledge. "The organizational choices are very localized and path dependent, throughout the game of organizational learning effects that they imply and by the specific investments, equipment or not that they mobilize". This is the reason why interaction between partners takes a unique feature.

This explains why the implementation of cooperations in the R-D field requires a sharing and a stable organization. As matter of fact, in the cooperation forms, the elements that will lead to the identification of cooperation will be: confidence and organizational learning as an output of this kind of relations.

3.2   About confidence

We can wonder what is the part played by trust whose importance, according to empirical surveys and theoretical works, makes it as a key component of cooperations and interindividual cooperations. Is this notion a necessary condition to justify the choice of a cooperation strategy? "How can it influence partnership"?

The notion of confidence is ambiguous because as an immaterial good it relies on the notion of reciprocity. If as LUNDVALL [14] says its material root is interdependency, it appears that the reciprocal aspect is present in the organizational learning, developing itself inside the interaction. Therefore, in a first part we have to put forward the confidence and learning links resulting from reciprocity and interdependencies. We propose, in a second part to underline their role in the cooperations. This allows to justify according to us the choice of a cooperation strategy.

Confidence as an active tool resulting from apprenticeship

In order to demonstrate how this specific asset is linked to learning, we must remind ourselves that confidence is not material and has no constraints. Assimilated to an implicit object non formalized between the actors, it is defined as the subjective attitude. Its role is to increment the quality of technological creations localizing at the same time apprenticeship and the know-how. Its feature of specific good comes from the particular relations between partners as a unique product of a relations requiring time. As a specific asset, confidence minimizes today and future transaction costs, but it especially has a reciprocity feature between partners [17] interest that we called organizational learning. However this last does not only imply a technical knowledge but also integrates a common know-how knowledge. This combination of knowledge has a reciprocity relation. Confidence, relying on reciprocity gives the opportunity to go on instead of the hypothesis according to which cooperation can be seen as a specific organization and coordination mode. As a matter of fact, confidence behaves as a self-reinforcement of the implementation of partnership. The interdependency of the actors, building new knowledge, generates at the same time knowledge and confidence. "Confidence is essential. It is not surprising to see that it is the cooperation tool. We note two main ways at the creation of cooperation: (1) to develop long-term relations, (2) to try to modify the game, acting on four variables (rewarding of cooperation, opportunism gains, punition and naive pay-off) keeping in mind the importance of the future" [9]. What first influences the choice of a cooperation strategy can be featured by reputation effects. This last compensates an ex-ante confidence failure, even its non existence. If we add the part of the learnings in the obtention of a dynamic organizational equilibrium, we can then put forward the hypothesis that the confidence linked to learning is one of the stabilizing elements of the choice of cooperation as strategy. If confidence is a specific asset resulting from the interaction between partners, the organizational learning allows to reinforce its development.

Trust and organizational learning as partnership bases

The tacit, irreversible and localized aspects of knowledge, allowing to encompass the difficulties met during the memorizing and development of learning processes reveal the importance of people's interactions. From then, coordination by cooperation is a conceptual frame that allows to analyze the objectives of the organizational learning, that is to say the developments of knowledge and confidence.

In this context, firms add various knowledge for the mergers. LORANGE and ROOS make the hypothesis that a partnership has for objective to gather competencies. One of the goal is then to "learn from an other partner how to realize a complex work" [21]. The key feature of success for a cooperation is in those conditions the creation of the organizational learning. Therefore, cooperation is not a stage towards the integration because actors make advantage of the fact to develop a distinct cooperation of their own organization. We propose in fact this causality to be inversed. The affirmation of an identity is done throughout the time, although the subject (firm or individual) changes. Moreover, if the identity is a social output, the cooperation claims the identity of partners inside its interaction. It inforces their belonging and their roots in the society. The functioning components of cooperation relies on the organizational flexibility and on its ability to develop common knowledge. The organizational learning of cooperation joins the building of habits and rules inside an organization, thus the ability to solve new problems produced by the instability of the environment and/or its behaviors towards other agents. The organizational flexibility is the ability to adapt itself to the unknown environment and the one to allow the organization to maintain its coherence. Organizations managed according to cooperation rules are dispositions faced to the various unknown sources and complexity.

4   CONCLUSION

As it introduced, concerning its specificity, cooperation is based on confidence rules, specific asset resulting from the interaction of partners and on the development of organizational learning. Cooperation organization is the product of actors' strategies in order to create a basis for the development of knowledge and learning. But the bilateral to multilateral field reinforces this conclusion in the way that the reciprocity system is more needed in the perturbed and unknown environments.

5 REFERENCES

[1] AMIT R. and SCHOMAKER P.J.H., Strategic assets and organizational rent, Strategic management Journal, vol 14, 1993, 33-46.

[2] COHENDET P. and LLERENA P.(eds), Flexibilité, information et décision, Economica, 1989.

[3] COHENDET P., HERAUD J.A. and ZUSCOVITCH E., Apprentissage technologique, réseaux économiques and appropriabilité des innovations, in FORAY D. et FREEMAN C. (eds), 1992, op. cit.

[4] DOSI G., TEECE D. and WINTER S., Les frontières des entreprises: vers une théorie de la cohérence de la grande entreprise, Revue d'Economie Industrielle, n^0 51, 1, 1990, p. 246.

[5] ELIASSON G., The firm as a competen team, Journal of Economic Behavior and Organization, 13, 1990, p. 276.

[6] FAVEREAU O., Organisation et Marché, Revue Française d'Economie, n^0 IV, 1 Hiver 1990, 65-148.

[7] FILIPPI M., Coopération industrielle et systèmes productifs localisés: une analyse en terme de réseaux, Thèse de Doctorat, Université de Corse, juillet, 1995.

[8] FORAY D., Repères pour une économie des organisations de Recherche-Développement, Revue d'Economie Politique, n^0 5, sept-oct 1991, 780-808.

[9] JARILLO J.C. and RICART J.E., Sustaining networks, Interfaces 17, sept-oct, 1987, p. 90.

[10] KIRAT T. and LE BAS C, La technologie comme actif, de la firme porte-feuille à la firme organisation, Revue Française d'Economie, vol VIII, 1, 1993, p. 155.

[11] KIRAT T., Pourquoi une théorie évolutionniste du changement technologique, Economie Appliquée, tome XLIV, n^0 3, 1991 p. 53.

[12] LANGLOIS R.N., Economic change and the boundaries of the firm, in Industrial Dynamics, Technological organizational and structural changes in industries and firms, in CARLSSON B. (ed), Kluwer Academic Publishers, 1989.

[13] LE BAS C., La firme et la nature de l'apprentissage, Economies et Sociétés, série W, n^0 1,1993, p. 20.

[14] LUNDVALD B-A (ed), National Systems of Innovation: towards a theory of innovation and interactive learning, Pinter Publishers, 1992.

[15] MEZIAS S.J. and GLYNN M.A., The free faces of corporate renewal, institution, revolution and evolution, Strategic Management Journal, vol 14,1993, p.78.

[16] NELSON R.R. and WINTER S., An evolutionary theory of economic change, The Belknap press, 1982.

[17] OUCHI W., Markets, bureaucraties and clans, Administrative Science Quaterly, vol 25, march, 1980, 129-141.

[18] PAULRE Economie et Société, série Dynamique technologie et organisation, W, n^0 1, 1993, p. 35

[19] PENROSE E., The Theory of the growth of the firm, Basil Blackwell, Oxford, 1959

[20] RICHARDSON G.B., The organization of industry, Economic Journal, septembrer 1972, 883-896.

[21] ROOS J., LORANGE P., Strategic alliances, formation implementation and evolution, Blackwell, 1992.

[22] RUMELT R., Theory strategy and entrepreuneurship, in TEECE D. (ed), The Competitive Challenge, 1987.

[23] ZUSCOVITCH E., LE BAS C., Apprentissage technologique et organisation: une analyse des configurations micro-économiques, Economies et Sociétés, série Dynamique technologique et organisation, W, n^0 1, 1993.

20

MULTIMEDIA EDUCATIONAL ISSUES IN A VARIETY OF LEARNING CONTEXTS

J-P. Soula*, C. Baron**

INSA-CCG, CERDIC
*** INSA-DGEI, LESIA*
Complexe Scientifique de Rangueil
31077 Toulouse Cedex FRANCE
e-mail: {baron, soula}@insa-tlse.fr
Phone : (+33) 05.61.55.94.90
Fax: (+33) 05 61.55.98.08

ABSTRACT

The 5th Advanced Technology Workshop (ATW) is a forum organized every year in both Europe and US. It represents a communication platform for participants from academia, government and industry to expose their activities relevant to the dissemination and application of new technologies, with an accent on practical solutions and an engineering curriculum reform. This year's European segment was hosted at the National Institute of Applied Sciences (INSA) of Toulouse, France, in July 1996; the topics particularly emphasized the commercialization of research, the quality and dependability of embedded systems, and collaborative engineering. Several panel discussions dealt with the question of efficiency of interaction between industry, government and academia, research today and tomorrow, research economics, and collaboration in research. A panel on *the use of multimedia and decentralized means of education* was organized in that context. Various experiments and projects were presented by teachers, researchers, governmental and industrial representatives coming from various countries (including France, Spain, UK, the Netherlands, USA, Hungary ..), addressing key educational issues. This paper presents a synthesis of what was exposed and discussed during this panel. We first review some experiments which aim at the integration of new technologies into learning strategies and which globally yield positive results. We then look at methods used by industry to develop educational software. Other experiments focused on the problem of integrated learning systems.

The list and addresses of panellist is given below. You are invited to contact either the panel coordinator Dr. JP. Soula for further information on the content of the paper, or Dr. C. Baron if you are interested in the ATW philosophy/participation.

Panellists Françoise BOISSIER
        DIAF Company, FRANCE.
Blaise J. DURANTE,
        Management Policy & Program Integration,
        Air Force Pentagon, Washington, USA.
        E-mail: blaise@af.pentagon.mil
Didier LAFRIQUE,
        Airbus Training, FRANCE.
Andrzej RUCINSKI,
        Dpt of Electrical & Computer Engineering,
        University of New Hampshire, Durham, USA.
        E-mail: andrzej.rucinski@unh.edu
Pablo P. SANCHEZ,
        Microelectronics Group, University of Cantabria, Santander, SPAIN.
        E-mail: sanchez@teisa.unican.es
Marilyn FILIPPI & Jean-Frangois SANTUCCI,
        Dpt. of Economics & Dpt. of Computer Engineering,
        University of Corsica, Corte, FRANCE.
        E-mail: santucci, mfilippi@lotus.univ-corse.fr
Jean-Pierre SOULA, coordinator,
        Management & Communication Center,
        INSA (National Institute for Applied Sciences), Toulouse,
        FRANCE.
        E-mail: soula@insa-tlse.fr

1   INTRODUCTION

The knowledge acquisition paradigm is undoubtedly being slowly transformed through the gradual introduction of new tools in learning environments. How this factor modifies the cognitive processes is not yet fully understood. It is to be observed that experiments have been launched in a variety of fields: information technology, cognitive psychology, language learning, physics, mathematics, etc... We might even venture to say that no body of knowledge is left untouched by what could be called the multimedia (MM) fever. This MM explosion can be accounted for by real needs in distance learning for instance to overcome geographical and/or economic difficulties (Canada and Australia have been pioneers in distance learning for example, for obvious geographical reasons) and perhaps at times by something akin to what might be called the 'jump on the bandwagon' syndrome.

Whatever the reasons may be we can say with Dr. A.Rucinski that there is a definite shift 'from chalk to virtual reality'. Indeed in a traditional educational environment the teacher-learner contact hours are the basis of transmission. Knowledge has been stocked on paper for many centuries and books are still the basis of learning even though audio, video tapes and floppies have been introduced together with virtual reality software which reproduce 'realistic laboratories'. Teachers have been using chalk, white board pens, transparencies, video recordings or computer programs as props for their job of knowledge transmission. Multimedia and decentralized means which are being introduced might tend to modify the standard learner-teacher relationship.

Toulouse's Advanced Technology Workshop (July 1996) organized a panel on the use of multimedia and decentralized means of education. Various experiments and projects were presented by researchers coming from various countries, addressing some educational aspects and problems. We will first review the experiments that integrate new technologies and which have a global impact. We will then look at methods used by industry to create educational software. Lastly, with other experiments the problem of integrated learning systems will be tackled.

2   'PROMISING RESULTS'

2.1   New Hampshire's Experience

Andrzej Rucinski (AR) presented the evolution of educational technologies "from chalk to virtual reality". He recalled the time when the use of transparencies was controversial, when the piece of chalk represented the teacher's insignia, though slides were very easy for preparing drawings for example. Next, television was introduced, mostly for teleconferencing applications; soon after video appeared, but according to AR, this means must be considered more as a tool of entertainment (nice to watch) to use when the student's attention decreases. Later, an attempt to bring personal computers into the classroom failed, because the instructor could not communicate with his audience: students spent their time on their keyboards, and were working on different topics, at various speeds. So, in AR's opinion, PCs could be useful for educational purposes, but their application should be dedicated to tutorials primarily. He finally contemplated multimedia as a cost-effective solution towards virtual reality.

However, he observed that the extensive use of chalk and transparencies in a classroom still exists in 1996. In his opinion, to thrive and develop a global infrastructure, the MM market needs to be more clearly identified; one of the things which could make the MM business feasible would be to provide learning on collaborative engineering. Moreover, for a successful use of MM as an educational means, the role of the teacher needs to be redefined: overwhelmed with the complexity of knowledge, he can no longer just be the symbol of knowledge, but a guide for students to identify and absorb knowledge on their own, through self instruction systems.

AR also mentioned his own experience in the use of MM for pedagogical applications. He observed that new technologies modify communication behaviours. Indeed, he personally faced the problem when he conducted a Virtual Classroom experiment between the University of New Hampshire (USA), and the Technical University of Budapest (Hungary) in Autumn 1993, using Internet, before the time when the World Wide Web was developed. His conclusions were that there must be back and forth movements, adjustments and fine-tuning, before new technologies can be totally integrated in an educational context. To comfort this opinion, he presently plans a MM-based lecture for the period of Autumn 1996 using standard videoconferencing procedures.

Concluding on his future plans concerning the use of MM, he envisages an evolution of MM use between the two European and American sessions of the next ATW exchange forum, videoconferencing for example their panel discussions so that American attendants can watch and interact with European panellists, and vice versa. He also referred to another benefit to be derived from decentralized communication which has been identified by the University of Corte. Dr. J-F Santucci sees it as a way for Corsica to break away from its geographical isolation as an island and improve inter-university cooperation - with New-Hampshire, Cantabria and Midi-Pyrénées, just to mention the panel participants - and to explore new cooperation channels between industry and academia.

2.2   US Army and Pentagon

With the reduction in human resources and funding, and with added educational requirements, the development of a distance learning (DL) capability became an imperative for the Air Force Institute of Technology (AFIT); in 1991 was thus created the Centre for Distance Education, whose mission was to develop an efficient means of exporting courses to meet AFIT's customer needs. They designed several multimedia training programs, which insured great success with DL, for example:

- the training centres use of MM for operational and maintenance training, for example, the use of 3-D engineering drawings with video to show how specific parts of a complex weapon system operate in various conditions

- the 81st training Support Squadron development of highly interactive computer-based training products using a full range of MM formats -texts, graphics, animations, photos, video, music, and narration- delivered on CD-ROM. Future training could be given through virtual classroom.

Dr. Blaise Durante (BD) also reported on an interesting experiment of implementation of interactive MM training programs for Surgical Instrumentation, at the National Naval Medical Centre of Maryland between September 1991 and June 1992, on a class (n^0 1) of 24 students. This was the first time the courseware was formally validated. The courseware was used for a significant number of hours and in a variety of ways; tests took place after each instructional session, and compared with the results of students who received standard instruction (n^0 2). The mean score for all tests was achieved by 98% of classmates 1, which represents a significant increase over the scores of class 2 (95%); moreover, out of the total number of test scores, 96% received A grades in class 1, whereas only 76% in class 2 received the same grades. These results indicated that the courseware was effective. In fact, both students and instructors generally agreed that it was a beneficial additional resource for teaching and learning (for example, videodisk images were preferred to the slides previously used), as it is not meant to totally replace student- teacher contact hours.

In BD's opinion, beyond the pedagogical advantages MM provides (real time updates, just in time knowledge, bringing the training to the personnel, collaborative education, multiple student and teacher interaction opportunities), their investments in MM and DL are economically worthwhile with regard to the gained benefits (cost per student declined by a factor of 10, possibility to train more people, more often, reduction of travel costs, flexibility to add students as needed without incurring additional expense). Indeed, they can for the most part benefit from the use of an up-to- date hardware equipment, and of already existing medias such as satellite (courses taught at satellite up-link stations can be broadcast to any base with satellite down-link capability, which every base has), Internet, or CD-ROM. As for software, the cost of developing MM products has been reduced significantly within the last couple of years because the creative processes & procedures are clearly mapped out as shown below; the commercial market has produced MM development tools that allowed the Air Force to develop better their own training products.

However, when making MM products, it is important to remember that students learn more through interactive processes, than passive processes. The products thus need to be user friendly, responsive and informative, in order to enhance the classroom by providing the instructors with innovative ways to present educational information.

To conclude this part, one key point about effective use of multimedia seems to rely on acquiring the talent of educators to create the educational material using the wide variety of available media*. Though some Air Force organizations can produce multimedia packages, such as the 81st training Support Squadron, there are many companies which provide this service at competitive prices because they have developed expertise and have fine tuned the cost aspects over several years.

3   METHODS AND METHODOLOGY TO DESIGN EDUCATIONAL SOFTWARE

The positive results recorded in MM learning depend on the global learning environment and no doubt on the quality of the software packages that are going to be the basis of data input. High standards of MM packages cannot be disconnected from economic concerns, as seen in the US Air Force case. We may assume that a fair compromise has to be found between graphic and pedagogical quality and economic issues. Two firms based in the Toulouse area have been devoting ressources and expertise to developing tools and processes. The first, DIAF is a firm specializing in various forms and fields of information technology. The second is the European Airbus consortium which has been a pioneer in computer assisted learning and computer assisted simulations.

3.1 The various trades in MM software production and cost-effectiveness.

DIAF with practically ten years experience in training software production, has produced packages for a variety of French national banks, for the aeronautical industry, for telecommunications and computer firms. They are currently involved in MM packages for the "Cité des Sciences et de l'Industrie" (Paris) which gives them expertise in museum software. Their challenge in this panel was to give an overview of the various trades to be found in MM production.

According to Françoise Boissier (FB) who is in charge of DIAF MM department, "Multimedia brings together text, pictures, sound and video within the same software to enhance an author's objectives whether they be game related, or for instruction, training or information purposes". FB identifies ten different types of activities or "trades" in the MM production field.

The author: he is an expert in his field or has the capacity to bring together experts of a given or related knowledge areas. The author may or may not know about computers and softwares but has to write out his ideas under a story board format. He is responsible for content and for pedagogical scenarios. It is obvious that his work with the other MM production team will influence his views on content presentation.

The interactive designer writes out a detailed story board -he may be the author himself- prepares specifications for the various components of MM production (computing, graphic, video and sound). He is the project manager as he must plan, look after costs and deadlines and make all the people involved in the project cooperate as smoothly and efficiently as possible.

The art director determines graphic styles and the leading presentation principles which involves close work with the video producer and the various graphic artists. This is also a team leader function.

The video producer prepares the scenario of video parts and he is in charge of filming them. He may have to look for archive pictures or films and to negotiate copyrights. Beyond the artistic function, he must also deal with post-production and editing work.

The graphic artist designs and creates screens together with 2D or 3D animations. If pictures have to be imported he has to scan them and touch them up. This artist also designs the graphic aspects of video clips.

The computer program developer (CPD) writes the program according to the specifications the interactive designer has produced. Incorporating graphic, animated and video parts falls within his province. The CPD has eventually to test the finished product before it is delivered to the customer.

The six former trades concern MM production in general. In the case of CD-Roms extra skills are necessary.

The producer will have a global view of the process as he determines the general outlines of the "title" - to use the trade jargon - which implies deciding on a title graphic style and the use of video. Some overlap with the interactive designer's and art director's areas might be observed. He looks after the financial aspects of the project and sees to the production's first tests and pressing of the CD-Rom.

The pressing specialist produces the CD master copy and is responsible for duplication.

With *the testers* we observe the recent development of a new specialized trade, triggered by the ever-increasing number of CD-Rom titles. The tester conducts limit tests of the CD master copy on all kinds of computer configurations

At the end of the line, *the distributor* has to sell the product and this is giving rise to a new type of distribution outlets.

As we can see, quite a number of skills beyond those directly related to computers, are involved in the design and making of an MM tool or a CD-Rom which accounts for the overall significant cost. F B gave the following figures to help us gauge how costs broke down for a CD-Rom that would correspond to an invoice for 1 to 3 million French Francs. Pre-production costs -including authors rights- correspond to 18% whereas copyrights might account for 15% -and in some cases for up to 35%- of the price.

Producing the CD-Rom proper requires about 55% of the total amount: 25% for graphic art, audiovisual and sound, 5% for digitizing and 25% for computer development. Creating the master copy involves 2% of the total amount and testing 10%. FB stressed the fact that these figures were given by way of global indication and that the share of the author seemed proportionally modest.

3.2 Airbus authoring workbench

The second industrial example, that of Airbus Industry shows a process similar to DIAF's general approach leading to the formalization of software design & development in a given context thanks to a specific authoring program. Once the training application area has been thoroughly mapped out, an "authoring workbench" is, according to Didier Lafrique (DL), a cost-effective solution.

Airbus Training's purpose is to develop simulation tools recreating the cockpit and the aircraft behaviour used to train pilots in order to put them in real-life situations. Their job also consists in maintaining the courseware, and updating it, because aircrafts are continuously improving. To do so, they use MM technologies to recreate the aircraft environment. DL's interest is more particularly focused on the methodology of MM tool production. As these tools are dedicated to the same application context (aircraft simulation), Airbus Training has implemented a tool providing software developers with a general behavioural architecture to derive the MM products.

Indeed, Airbus Training involvement in the development process of MM application is double since catering for pedagogical efficiency, they try to reduce MM development costs.

To meet the first point, they take advantage of Windows standard MM features to enhance the operational aspect of courseware (realistic cockpit panels, morphism, video clip, etc ...).

As cost effectiveness is also very important, two concepts were introduced in their original authoring tool to make the development and modification of courseware easier:

(i) every time an element or a function is often used in lessons, this feature is pre-programmed at the workbench level. This characteristic not only improves the courseware quality in that we can be sure that the behaviour would be the same within the courseware, but it is also beneficial to facilitate the standardisation (and the re-use) of it.

(ii) the tool *modularity* facilitates the incorporation of new functions.

This kind of features enhance the technical accuracy of the courseware, but also improve development and management aspects.

A last element to be mentioned is the introduction, at the top of the "develop-ment edifice", of a module maker, which allows to develop in a wysiwyg way: the developer can see exactly what the trainee will see at the end. And again, this enables the producer to speed up the courseware development phase and so be more cost-effective as Blaise Durante pointed out in the case of the US Air Force products.

4   ALL THAT GLITTERS IS NOT MM

4.1   The multimedia lecture: educational revolution or new software business?

Dr. Pablo Sanchez (PS) presented his first views concerning 2 experiments started in July 95 and scheduled until June 98. This project groups together 3 computer science departments of the University of Cantabria (Spain).

The objective of the first project (CICYT-TIC 95-0837-C02) is to give stu-dents an MM package helping them to learn about microprocessors through a simulation-based approach. The various tools used are a program editor, a debugger, a logic analyser and simulator and an MM user interface. As far as the second experiment is concerned, it aims at giving students CAD training with a software developed with Asymetrix Toolbook 3.0. PS has no figures to propose at this stage but he has been able to identify two sets of problems. The first series belongs to the technological field. A computer, as a tool still has limitations in spite of the tremendous headway made by technology. For instance, depending on the machine being used, graphics may be rather poor in quality and voice and image recognition is far from being perfect. Video generation may be another problem area and so can be speed. Turning now to software, recognition algorithms can be found fault with, Artificial Intelligence programs may at times suggest that for intelligence in A.I. is something of a misnomer. Lastly, communication is not always easy to establish between dif-ferent programs. However, these grey areas can be expected to be cleared in the near future.

PS lays special emphasis on pedagogical problems. He focuses more specifically on learning processes and wonders whether it may be possible to propose a learning model. He raises the problem of information flow in quantitative terms, asking if it is possible to establish values that could help optimize it. Beyond this quantitative view he expressed qualitative concerns too, suggesting that 'conceptual doubt' as he calls it - viz. making the student think about what she/he is learning and thus raise questions about it- can be a step in the learning process. To PS's mind, a software package should place the learner in a position where he/she will be led not only to ingest data but to hesitate and activate his/her doubting potential. In other words, the learning process is not only connected to a quantitative intake of data but it calls upon deeper thinking processes that can help assemble data through active questioning. Clearly, this view which indicates that a student must not be passive but on the contrary actively involved in the learning process, bridges the gap with cognitive sciences concerns and echoes issues raised by Rucinsky and Durante.

At the present stage of the on-going teaching experiments of the University of Cantabria, PS is tempted to suggest two opinions to answer the question raised in the title of his paper. A multimedia package that would provide data on a given topic in an attractive format, that would give a data base and other tools meant to provoke thinking -or "doubt" as PS says- would have a high impact as far as education, training are concerned. If on the other hand a MM package has only the pleasant presentation of data as a major feature, it can perhaps replace books but will not be more efficient from the learning point of view. This second type of software packages is no doubt easier to produce if authoring packages are used once the target public has been identified and they can have a high social impact and be good for business. In other words, to the initial question 'is the MM lecture an educational revolution or a new software business?' Pablo Sanchez states that the second proposal, to his mind, is the correct answer.

4.2   The screen's the limit

Dr. Jean-Pierre Soula (J-P S) similarly tended to be cautious when it came to the use and social impact of MM tools. His caution stems from the fact that quite a number of decision-makers in education or industry have focused their attention on purely economic issues, repeating errors and misconceptions which had been made some twenty years ago with language laboratories. It was thought then that language labs were going to miraculously solve the problems of language learning and this has yet to be proved! Many decision-makers have been entertaining the dream that computer ressource centres or labs could significantly replace language teachers and so be a cost-effective solution. Past and on-going experiments at INSA with engineering students seem to suggest that this view is overoptimistic.

The first experiment consisted in giving access to the traditional language lab to students who volunteered to brush up and/or expand their English, Spanish or German using self-study audio packages. In some examples when motivation was extremey high, results have been astounding. Unfortunately these cases are few and far between. Over the several years (eight) since this self- access possibility has been given to students, global interest and motivation appear to have waned. One of the explanations put forward is that students miss human interaction.

The second experiment with autonomous learning on a self-study basis implied MM tools. When setting up INSA experimental MM language ressource centre it was decided to observe users for the whole academic year. Users could come for 3 reasons; (i) they had an assignment which required the use of data to be found on a CD-Rom or through Internet, (ii) they were part of a scheme of guided autonomy implying meeting a teacher at appointed times over a three-month period, (iii) they could come and work on their own as free agents and do any type of linguistic work they wished. Relatively few students came on a regular basis for the last reason which confirmed what we had been previously observed for the language lab.

J-P S indicated however that many aspects of language learning can be acti-vated thanks to MM packages provided the use of CD-Roms is part and parcel of a global scheme. In that case, the autonomous and individual work with a computer to revise or store linguistic data will become meaningful if oppor-tunities are given to use them, on site, in realistic contexts with peers and/or a linguist. Interfacing with a machine is not enough, as communication im-plies much more than the purely verbal data. The role of gestures, intonation, attitudes, eye and hand movements, in a word affective/emotional clues in a communication process must not be overlooked. It had been thought that the sky was the limit concerning MM language learning but at INSA they tend to think that" the screen's the limit"!

J-P S believes that MM tools cannot be self-sufficient but that they can be powerful adjuncts if included in an overall language learning approach that will not forget the human factor. Such combinations can be implemented on a given site and/or with other sites abroad. If funded, a European project -within the Leonardo framework- including the Netherlands, Great Britain, Spain and France will produce simulations that will allow students from the 4 countries to use the languages of these countries in order to solve cases and problems on a variety of company or society-related topics. MM tools will be used on site to present cases and groups of students will interact with each other first within their own community (in their University) and then they will interact via Internet - E-mail and low-definition videoconferencing- to deal with their case.

This corresponds to simulated work within a purely academic context, but the same principles of decentralized cooperative work are presented in Santucci's and Filippi's project for the University of Corte. They aim at fostering coopera-tion between Industry, Administration and Academia. Cross-disciplinary work between economists and computer scientists at the University suggested that these collaborative principles could be modelized. On-site collaboration and off-site meetings through video-conferencing could shape a new methodology for collaborative work.

5 CONCLUSION

Reviewing these various experiments and projects revealed that comparable findings have been reached in the different cases in spite of the diversity of experimental fields. The outcomes may not be exactly identical because the environments in which MM learning tools were used differed on several counts. Knowledge transmission culture is bound to be different in Spain, in the United States, and within that country, academia and the army have cultures of their own, in the same way as in France, academia and industry follow rules of their own. Beyond the differences in national and functional cultures the vari-ous panellists referred to different academic fields; computer sciences were well represented but medicine, economics and foreign languages had the floor too. Whatever the fields and/or countries four types of questions referring to cog-nition in general, have been identified (i) the tools themselves (ii) the human factors (iii) cost-effectiveness and MM and (iv) systemic approach.

(i) *MM tools.*

Software packages should be thought-provoking and enable learners to experi-ence "doubt" as Pablo Sanchez put it. If MM tools are only glorified books they may be pleasant to look at but they will short-change users. It then follows that research work could be done in this area. How can MM computers be put to best use to facilitate learning processes? How can the current interactive possibilities be improved upon to trigger cognitive processes?

Another line of enquiry concerns MM authoring packages. In two cases, ref-erences were made to Toolbook by Asymetrix and we might thus wonder to what extent a given tool influences the way authors and CD-ROM producers organize their content presentation and interactivity scenarios. Experiments in this area should shed some light on creation strategies and processes.

(ii) *The human factor.*

Machines cannot eliminate learner/teacher contacts because in spite of A.I. advances expertise cannot be totally modelized. However, as Andrzej Rucinski pointed out, there will be a transformation in the educational teacher/student relationship. The role of the teacher is likely to gradually alter from the lecturer's omniscient position to that of a consultant or expert in a given area who will have a facilitating role to play. How is this relationship going to be modified if learners and experts communicate through the net for consultations and if videoconferencing is used to decentralize knowledge transmission? What is going to change in a University if the standard learner/teacher relationship becomes learner/CD-ROM/teacher? We can indeed by trial and error endeavour to find viable local models and this will undoubtedly happen. Should not we however take this as a serious question and devote resources to that type of applied research where cooperative projects could be run by information technology experts and cognitive sciences specialists? By way of illustration let us mention a research project conducted in the Toulouse area in the early days of TV videoconferencing. The telecommunication specialists wondered what could happen in a negotiation context when the different parties were miles apart. Experiments were set up with a Toulouse psychology laboratory and behaviour change patterns were identified. In a MM educational relationship, we are in a similar position.

(iii) *Cost-effectiveness and MM.*

MM can foster financial gains as Durante observed in connection with the various USAF schemes, since travel time and expenses could be dramatically reduced and leave some financial benefits after computers and software had been bought and tutorial and follow-up procedures organized. However, from a more general standpoint, we wonder whether it might not be safer to globally consider that with similar expenses, better results can be obtained when integrating MM in a system - even though it may be true that in some instances financial savings can be immediately observed. Education and training had better take a long-term view on instruction and think in terms of global systems.

DIAF and Airbus clearly demonstrated that initial investments to analyse MM production processes and their application environments could be offset by the derived expertise which in turn helped bring costs down. Beyond this directly financial outcome we may wonder whether we cannot also find a cognitive cost-effectiveness. If educational softwares are not produced in an empirical way -which is often the case for handbooks- but integrate cognitive concerns similar to those identified by Pablo Sanchez, then we might assume that they could help save on learning time, which would be both beneficial to individuals and to institutions alike. This cognitive benefit is connected to an overall systemic view as suggested by the Filippi/Santucci project.

(iv) *Systemic approach*

It can be inferred from all the panellists' accounts that MM deserves better than being artificially grafted onto an already existing instructional system. To start with, it was necessary to experiment with MM in a variety of contexts to bring to light its assets and liabilities. MM instruction can be a system in its own right when specific geographical conditions exist but the human factor cannot be disregarded, no more than it can be naively considered that the same learning/teaching paradigms as in on-site education will apply. On the other hand, when introducing MM in an already existing educational system - universities, colleges and companies alike- one should carefully assess what will be transformed in the system, so that real integration or embedding can take place. Careful planning is required, for the advantages of MM learning to be added to the prevailing advantages.

In a nutshell, MM instruction should not be viewed as an end in itself but rather as a magnificent opportunity for all the actors - teachers, students, administrators...- of a given educational community, to take a fresh look at knowledge and at learning issues. Cross-disciplinary action implying fundamental and applied research may then be necessary.

SURF-2 A TOOL FOR DEPENDABILITY MODELING AND EVALUATION

Presented by L. Blain, J-E. Doucet

LAAS-CNRS 7, Avenue du Colonel Roche
31077 TOULOUSE CEDEX - FRANCE

ABSTRACT

SURF-2 is a software tool for evaluating system dependability. It is especially designed for an evaluation-based system design approach in which multiple design solutions need to be compared from the dependability viewpoint. System behavior may be modeled either by Markov chains or by generalized stochastic Petri nets. The tool supports the evaluation of different measures of dependability, including pointwise measures, asymptotic measures, mean sojourn times and, by superposing a reward structure on the behavior model, reward measures such as expected performance or cost.

1 INTRODUCTION

SURF-2 is a software toolbox that provides a working environment to support an evaluation-based dependable system design approach. It provides assistance for the construction of models of system behavior and the processing of these models to obtain various system dependability measures. Moreover, SURF-2 provides explicit support for carrying out comparative evaluations of multiple potential design solutions. The SURF-2 main principles are described hereafter. Generally, two main phases can be considered when carrying out a quantitative evaluation of system dependability:

a) construction of a model of the system's behavior from the elementary stochastic processes which describe the behavior of the system's components and their interactions;

b) processing of the model in order to obtain expressions or values of the dependability measures of the system.

Concerning phase (a), over the past years, stochastic Petri nets have emerged as a privileged approach for building Markov chains from the system components' behavior and interactions when those interactions imply stochastic dependencies. With SURF-2, a user can either (i) describe the model behavior by means of generalized stochastic Petri nets (GSPNs, see e.g., [1]) whose reachability graph is then transformed into a Markov chain, or, (ii) enter directly a Markov chain. Moreover, in the case of GSPNs, SURF-2 enables the models to be verified in terms of their structural properties. This allows the user to gain confidence in the models on which judgments about system dependability will be based.

Phase (b) above is carried out completely automatically by SURF-2. Measures of dependability for a given system are obtained by processing a Markov chain description of the system's behavior.

SURF-2 processes Markov chains that are either entered directly by the user, or derived from a GSPN model.

Three actions are carried out during this transformation:

- parameterization: variables in basic (generic) Petri nets are instantiated (e.g., the initial marking variables m and n appearing in Figure 1);

- structural verifications: they enable properties (boundedness, liveness or pseudo-liveness, as well as specific assertions on the markings) to be verified on the Petri net;

- computation of the reachability graph generated by the initial marking considered.

Figure 1 gives a single-model view of SURF-2. However, SURF-2 also provides support for considering altogether several models (i.e., Markov chain and GSPN models) in order to carry out comparative evaluations of competing solutions.

2   SURF-2 OBJECTS AND TOOLS

Figure 2 shows the structure of SURF-2 from the user viewpoint. SURF-2 may be viewed as a toolbox where each tool can operate independently on objects gathered in a database. The four main object types created by the user and

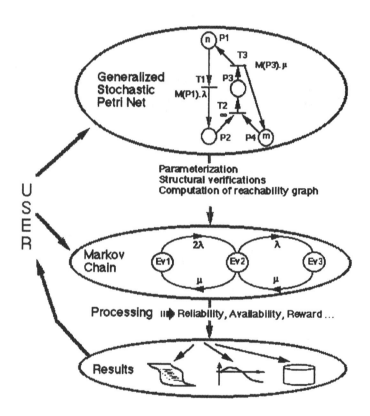

Figure 1   SURF-2 modeling framework

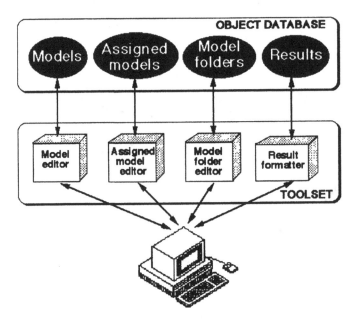

Figure 2   Structure of SURF-2 from user viewpoint

managed by SURF-2 tools are models, assigned models, model folders and results. Each object in the database is referenced by its name. All objects of a given type are gathered into a library for objects of that type.

SURF-2 is implemented as a multi-tasking package: one or more instantiations of each tool can be executed concurrently. Selection and synchronization of tools are managed by the supervisor, which runs as a permanent background process. The supervisor provides a menu windows interface to the user. A specific menu window is provided for manipulating each of the four main SURF-2 objects.

The flexible organization of SURF-2 as a collection of tools communicating through objects stored as files in a database proved to be a wise design decision since new tools can be easily integrated into the toolbox. Several system functions are in fact also implemented as tools (these tools cannot be used directly by the user).

A User's Guide documents in detail the main characteristics of the various tools available within SURF-2 and illustrates, by means of didactic examples, their usage and their capabilities [2].

2.1 Models

A model object consists of the description of the topology of a Markov chain or a GSPN that describes the behavior of a system. In addition to the topology of the behavior description, the model object contains the specification of the initial probability vector of the Markov chain (or the underlying Markov chain in the case of a GSPN description) and of one or more state partitions.

State partitions are necessary to specify how the dependability of the system is to be measured. Several state partitions can be defined in a given model. A state partition generally consists of two classes "called proper and improper" due to the principle of proper and improper service that underlies the definition of most dependability measures. For instance, the pointwise availability of a system, $A(t)$, is defined as the probability of the system state being in the proper (service) class at time t. An exception occurs for safety-type measures, which need the definition of an extra class "called catastrophic" to denote the subset of states of the improper class that represent a catastrophic (as opposed to benign) system failure. For instance, the pointwise safety of a system, $S(t)$, is defined as the probability of the system state not entering the catastrophic (failure) class before time t.

Parameters in model objects are specified either numerically or symbolically. A symbolic parameter is a purely local variable whose scope is limited to that of the model in which it is defined. Symbolic parameters, together with the possibility of defining several state partitions, allow the definition of very generic, reusable models that can be stored in a model library in the database.

Figure 3 illustrates the interface provided by the GSPN model editor (the Markov chain Editor is similar). Besides several general functions for manipulating graphical objects (delete, move, copy, etc.), these tools are true syntactic editors with features and operations specific to the type of models they support. For example timed transitions are represented by white "boxes", while instantaneous transitions are shown in gray.

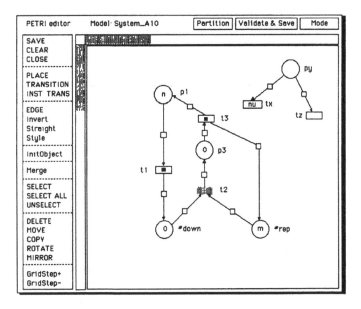

Figure 3   Generalized stochastic Petri net editor

The stochastic rates associated with the transitions are defined either by values, variables (e.g., rate nu for transition tx) or by formal expressions (such as "m(p1) * la", for example, for transition t1 where "m(p1)" denotes the marking of place p1). These definitions can be specified through specific editing windows obtained by clicking on the considered transition. If the definition of a transition does not fully fit within the transition "box" shown on the screen, then simply a black dot appears on the "box" (e.g., as shown on transitions t1 and t3).

In both cases (GSPN or Markov chain editors), although incomplete work can be saved, it is not possible to use a model whose syntax has not been validated. This is the case for the GSPN shown on Figure 3; the net description is incomplete: the subnets are not connected and concerning the subnet on the top right corner, the rate associated with transition tz is not yet defined.

2.2 Assigned Models

An assigned model object consists of a model and a model assignment, which is a mapping between the symbolic parameters of an object and a set of numerical values and global variables. This is done by the model assignment editor. All parameters of the specified model are automatically listed and the user enters either a value or the name of a global variable into each corresponding field. The latter possibility is provided to facilitate assignment of numerical values that are common to several assigned models, corresponding to the same or to different systems.

2.3 Model Folders

Comparative assessment of competing design solutions requires the simultaneous evaluation of a common dependability measure for several systems or system configurations. Such comparisons are the role of the model folder object. A model folder consists of a set of assigned models, a specification of the desired dependability measure and of the state partitions for each model that corresponds to the chosen dependability measure. The model folder also contains a global variable assignment, which maps numerical values onto the global variables defined in the set of included assigned models. The global variable assignment also allows several values to be mapped onto a given global variable. This feature is a convenient way of defining sensitivity analyses.

These powerful features are illustrated by the two deliberately simple examples shown in Figure 4.

In Figure 4-a, two models are to be compared. Model_1 and Model_2 both have three symbolic parameters. The same identifier, b, happens to be used as a symbolic parameter in both models. However, since the scope of symbolic parameters is limited to the model in which they are defined, this identifier can be mapped onto different values in the corresponding assigned models. Conversely, two distinct parameters have been defined in terms of a global variable $lambda to which a single value is assigned in Model_folder_A.

In Figure 4-b, two assigned models are derived from a common model. The variable m is given a different value in each assigned model, whereas l is defined in terms of the global variable $lbda. In Model_folder_B, the global variable $lbda is assigned a set of three values; consequently, both assigned models will be evaluated for three distinct values of $lbda. It is worth noting that the two values of m could likewise have been defined by an intermediate global variable defined as a set of two values in the model folder.

Depending on the measure to be computed some other global variables may appear in a model folder and must be instantiated by the user. For instance, pointwise measures involve the assignment of the global variable time (e.g. Reliability in Model_folder_b).

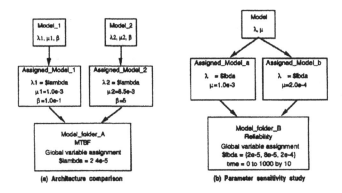

Figure 4   Illustration of the use of global variables

The model folder editor enables the user to select the set of assigned models (e.g. Assigned_Model_1 and Assigned_Model_2 in Model_folder_a), the dependability measure to be computed (e.g. MTBF in Model_folder_a and Reliability in Model_folder_b), the relevant state partitions for each included model, and to assign the global variables (e.g. $lbda and time in Model_folder_b). The user may choose among six pointwise measures, four asymptotic measures and seven mean value measures, as indicated in Table 1.

Since SURF-2 can compute reward measures, it actually has a much wider scope of application than dependability evaluation. Measures of expected performance, benefit and cost can easily be defined by means of appropriate reward structures (both reward rates and bonuses) mapped onto behavioral models [3]; the classical mapping onto Markov chain states and edges has been extended to the case of GSPN markings and transitions.

Table 1: SURF-2 measures

POINTWISE MEASURES	ASYMPTOTIC MEASURES	MEAN VALUES
Reliability	Availability	MTFF
Safety	Unavailability	MTTF
Maintainability	Reward	MUT
Availability	Reward rate	MDT
Unavailability		MTBF
Reward		AC
		Reward

2.4 Results

The last major object type held in the database are result objects. A result object consists of the set of numerical values obtained by processing all the assigned models in a model folder for all values of the global variables defined in the latter. Result objects may be displayed as a set of curves or tables by means of the result formatter. This tool provides a high degree of flexibility. A subset of the results obtained from a single model folder can be selected and displayed in a user-selectable format.

Figure 5 shows an example of the curve-plotter display offered by the result formatter. Results can also be displayed as tables of numerical values.

All objects handled by SURF-2 can be displayed on the screen or printed by means of intermediate PostScript files.

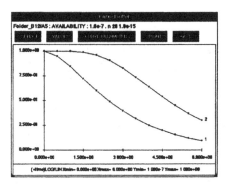

Figure 5   Curve-plotter display

3   SUMMARY

SURF-2 is a working environment for dependability evaluation. From the user viewpoint, the most significant features of SURF-2 are:

- a highly interactive menu-driven graphical user interface with powerful dedicated syntactic editors;

- an integrated environment for managing dependability projects that require comparison of competing design solutions or analysis of parameter sensitivity;

- the ability to calculate numerous measures of dependability;

- the ability to calculate reward measures such as expected performance or cost, by superposing a reward structure onto a behavioral model.
  SURF-2 has been implemented for Sun workstations running Solaris 1 and 2, and X- Window R5 graphic environment.

There is no hard-coded limitation to the size of the models that can be created and evaluated. To estimate the capabilities of SURF-2, several test models of several tens of thousand states have been processed on a Sparcstation 5 with 32 Mbytes of memory and a swap zone of 64 Mbytes. For such a configuration, the maximum size for a Markov chain is 150,000 states. SURF-2 has been used to evaluate the dependability of several real-life complex systems in different domains such as Air Traffic Control [4][5] and Electronic Automotive Systems [6].

4 REFERENCES

[1] A. Marsan, G. Balbo and G. Conte, "A Class of Generalized Stochastic Petri Nets for the Performance Analysis of Multiprocessor Systems", ACM Trans. Computer Systems, 2 (2), pp.93-122, May 1984.

[3] R. A. Howard, Dynamic Probabilistic Systems, vol. II, Wiley, New York, 1971.

[2] S. Bachman et al., SURF-2 User's Guide, LAAS-CNRS, Research Report, June 1996.

[4] K. Kanoun and M. Borrel, "Dependability of Fault-tolerant Systems Explicit Modeling of the Interactions Between Hardware and Software Components", IEEE International Computer Performance and Dependability Symposium, (Urbana-Champaign, IL, USA), September 1996.

[5] K. Kanoun, M. Borrel, T. Moreteveille and A. Peytavin, "Modeling the Dependability of CAUTRA, a Subset of the French Air Traffic Control System", in 26th Int. Symp. Fault- Tolerant Computing (FTCS-26), (Sendai, Japan), LAAS-Report 95-515.

[6] C. Ziegler, "Sûreté de fonctionnement des systèmes électroniques embarqués sur automobile", Thèse de Doctorat de l'INPT, July 1996 (in French).

* SURF-2 is the result of a collective work carried out by M. Aguéra, J. Arlat, S. Bachmann, L. Blain, C. Béounes, C. Bourdeau, J.-E. Doucet, K. Kanoun, J.-C. Laprie, S. Metge, D. Powell, P. Spiesser from LAAS, and J. Moreira de Souza from CPqD - TELEBRAS (Brazil). Christian Béounes passed away in 1993 but still lives in our memory.

22

SUBMICRON CIRCUIT DESIGN ON PC

E. Sicard, S. Delmas

LESIA-INSA, Department of Electrical and Computer Engineering
Av de Rangueil 31077 Toulouse Cedex 4 (France)
Fax: +33 5 61 55 98 00 - e-mail: etienne@dge.insa-tlse.fr
WEB: http:/www.insa-tlse.fr/ etienne

ABSTRACT

This paper describes a set of software tools well suitable for teaching graduated and undergraduated students the design of VLSI submicron integrated circuits. The CAD system *Introduction to Microelectronics* includes basic tools such as mask-level editor, on-line Design Rule Checker, electrical parameter extractor and fast analog simulator. The software includes also a tutorial on MOS and bipolar devices with comparison between simulations and measurements and a CMOS process simulator.

1 INTRODUCTION

In the last fifteen years, micro-electronics has known an important evolution and the growth in this field has been vertiginous (30 % for 1995). The Moore law states that the amount of transistors in micro-processors doubles every two years and it has always been respected up to day. MOS transistors have been scaled down in dimensions both vertically and horizontally. Unfortunately not all device parameters can be scaled proportionally. These limits have increased the importance of device and circuits modeling.

The past several years have seen an increasing need of engineers who have learnt about design techniques of integrated circuits and today this trend continues to be proved. After a brief description of the various MOS transistor models (section 2), the paper presents a set of home made tools available for PC environment. The software called *Introduction to Microelectronics* is currently used in 300 universities in Europe and in the world for teaching micro-electronics basics.

The two first lesson sets include the MOS theory and its integration. A behavioral illustration of the n-channel and p-channel MOS devices called *PROF* and a 3-dimentionnal CMOS process simulation called *3D* are detailed in section 3. The other lessons are divided into two sets, one dedicated to logic circuit design and the other one dedicated to analog design. Students use a layout editor and built-in analog simulator called *MSK* to mater both parts. The corresponding software is presented in section 4 followed by general conclusions (section 5).

2   THE MOS MODELS

Several MOS models have been developed in parallel with the technology evolution, as new phenomena and limitation factors rised in importance with the scaling down of the lithography [1]. Most analog simulators, such as the popular SPICE feature a wide range of models called levels. The *level 1* is defined according to the ideal transistor theory based on the Schichman-Hodge equations. This model is only available for long and large channels ($W > 10\mu m, L > 10\mu m$). The *level 2* includes more detailed device physics. This is not a version accurate for small geometry devices. The *Level 3* pursues a semi-empirical approach to model some short channel effects, but does not fit well with submicron technologies needs because its accuracy relies on the choice of empirical parameters.

The *Level 4* (also called *BSIM1*) was introduced in SPICE in 1983 by Berkeley Short Channel MOS Model. This model is used for channel length above 1 μm but it handles a large number of length and width dependent parameters which lead to complex extraction procedure. The *BSIM2* version, developed in 1989, fits devices down to 0.5 μm. But while the general formulation of this model is still physics based, many parameters have been introduced to increase computational efficiency. The very large number of empirical parameters are hardly implemented. The *BSIM3* version was proposed in 1994 [2]. It's a physical model based on a quasi 2-dimentionnal analysis of MOS device structure, taking into account the effects of device geometry and process parameters such as short channel effects and high field effects. The *BSIM3 release 3* was finally developed in march 1996.

The *MM9* model is a compact MOS-transistor model, intended for the simulation of circuit behavior with emphasis on analog applications [3]. The model gives a complete description of all transistor-action-related quantities. The equations describing these quantities are based on the gradual-channel approximation with a number of first-order corrections for small-size effects. The consistency is maintained by using the same carrier-density and electrical-field expressions in the calculation of all model quantities. The continuity of the derivatives of currents and charges has been a point of special emphasis. Especially the description of the transition from weak to strong inversion and also of the transition from linear to saturation have been thoroughly investigated.

3   TUTORIAL ON MOS DEVICES AND FABRICATION

When trying to provide the students with an understanding of integrated circuits, we generally introduce basics on MOS devices [4].

- The PROF program displays the characteristics of nMOS and pMOS transistors as well as NPN, PNP devices and diodes. The tutorial on n-channel transistor (the default device) is presented in figure1.

The screen is divided into four parts:

The process simulation window, situated near the left hand side upper corner of the screen, represents the vertical aspect of the device.

At the right hand side of the process window appears the bird's view of the device.

Situated below the process simulation window is the control panel. The students acts on the voltage value of the drain, the source or the gate, using either the mouse or the keyboard to move the cursor positions. Other information available in this control panel are the current value between drain and source (Ids) and the function mode.

The last window contains the MOS characteristics, including the operating point mark represented with an I . The channel aspect, the operating point location as well as Ids and mode indicators change each time the student acts on the cursors. All transistor parameters, such as the width and length, power voltage, and technological parameters can be modified. Both depleted and enhanced transistors can be simulated.

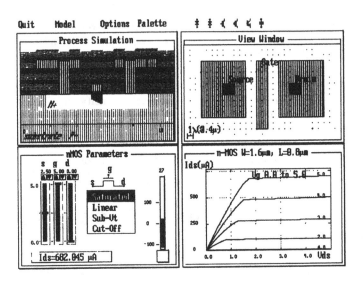

Figure 1   Tutorial on n-channel MOS devices.

In this tool, three models are provided to compute the MOS caracteristics: the MOS model 1, the model 3, and MM9. Moreover, the simulation may be compared with measurements.

- The 3D software tool provides to the student a comprehensive introduction to the MOS process techniques, with a simulation of the process fabrication. Polysilicon gate, self aligning CMOS process is supported, as well as Bipolar CMOS process. The student reads a mask-level layout and conducts the MOS fabrication step by step. The representation is based on a simplified but attractive three-dimensional simulation of the circuit using the entire color palette available on the graphic colour card.

- Advanced training in CMOS design should include the theoretical presentation of MOS long-channel, short-channel and second order effects, together with practical training where the student learns the role of each parameters in order to fit simulation with the measurements. Such a practical approach is proposed in PROF.

The program scans the current directly and displays the list of measurement files corresponding to the measures performed on a real MOS devices with different sizes. The measurements concern ATMEL ES2 1.2, 1.0, 0.7 μm technology. The measured data are added to the simulated result. The

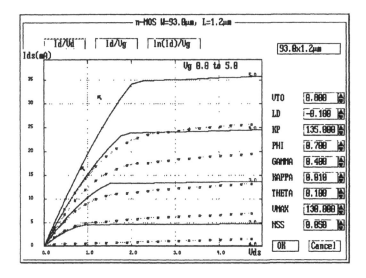

Figure 2 Comparison between simulation based on typical parameters and a measurement in a 1.2 μm process

students change the MOS parameters values directly on the screen and observe the effects of the fitting procedure. A comparison between default foundry parameters and measurements on a n-channel device is reported in figure 2.

The student acts on the MOS model parameters following a step-to-step procedure to tune the simulation to the measurements: threshold voltage, beta, mobility limitation, lateral diffusion, sub-threshold domain. Once the DC parameters have been extracted, the basic AC parameters are tuned using C/V measurements of a set of capacitances

4   THE LAYOUT EDITOR

Students have been given some CAD approach to MOS simulation using PROF and 3D modules. The layout design system called MSK features a similar user interface as for those modules [5].

■ MSK allows the student to design and simulate an integrated circuit. The package itself contains library of common logic and analog ICs to view and simulate. MSK includes all the commands for a mask editor as well as new original tools never include before in an unique module. The student can gain access to Circuit Simulation at the press of one single key. Electric extraction of the circuit is automatically performed and the analog simulator produces immediately voltage and current curves.

■ The Process Simulator shows the layout in a vertical perspective, as when fabrication has been completed. The Logic Cell Compiler is a particularly sophisticated tool enabling automatic design of a CMOS circuit corresponding to your logic description. The cell is created in compliance with environment, design rules and fabrication specifications. A set of standard 2-metal CMOS processes are available as well as advanced 5-metal deep submicron technologies, and BiCMOS process technology which enables the design of Bipolar devices with minimum technological changes.

■ Moreover, MSK allows comparisons between dynamic simulations and dynamic measurements [6]: a ring oscillator based on an odd number of inverters is proposed to characterise the delay due to interconnects. The layout is shown in figure 3. Considering the line models, the simple capacitance-to-ground model is used for CMOS technologies with 1 μm devices. The switching speed is low enough to consider the line as a simple discrete capacitance. A measured propagation signal in an inverter chain obtained by sampling electron beam measurement is given to the student who acts on the linear capacitance (Cmetal/substrat) value to fit the simulation with the measure as illustrated below.

■ The software handles full BiCMOS process (both NPN and PNP). Specific layers representing the N-doped base and P-doped base are used to create the vertical P+/N/P- and N+/P/N- bipolar devices.

Moreover, the five layers of metal are supported. In the cell library, the Metal2/Metal3, Metal3/Metal4 and Metal4/Metal5 contacts are available.

5   CONCLUSION

A PC-based software illustrating the principles of submicron VLSI design has been presented. It features a valuable introduction to the single MOS device, its models compared to its real behaviour, the CMOS technology, and the design of logic/analog cells.

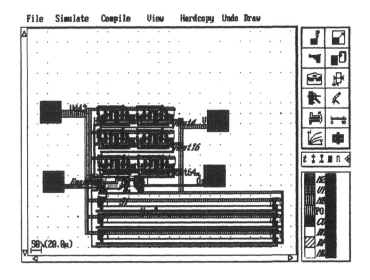

Figure 3 Layout of the ring oscillator used for comparison between simulation and real case measurements.

The Windows version of the software is under development. Very complex circuits designs will be supported, with almost no limitation in term of devices.

6 REFERENCES

[1] M. Napieralska, E. Sicard, S. Delmas, "Submicron MOS modeling: A compromise between accuracy and complexity", Mixed Design of Integrated Circuits and Systems, Lodz, June 1996.

[2] J.Huang, Z. Liu, M. Jeng, K. Hui, M. Chan, P. Ko, C. Hu, "BSIM3 Manual, version 2.0", University of California, Berkeley, March, 1994.

[3] R. Velghe, D. Klaassen, F. Klaassen, "MOS model 9", Report NL-UR 003/94, Philipps Research Laboratories, June 1995.

[4] E. Sicard, A. Rubio, K. Kinoshita, "A VLSI design system for teaching introduction to microelectronics", IEEE Transactions on Education, Vol 35, No. 4, November 1992, pp 311-320.

[5] E. Sicard, "Introduction to Microelectronics Version 5.1", INSA Toulouse Publishing, 1996.

[6] S. Delmas, E. Sicard, M. Napieralska, "New challenges in deep submicron interconnect modeling", Mixed Design of Integrated Circuits and Systems, Lodz, June 1996.